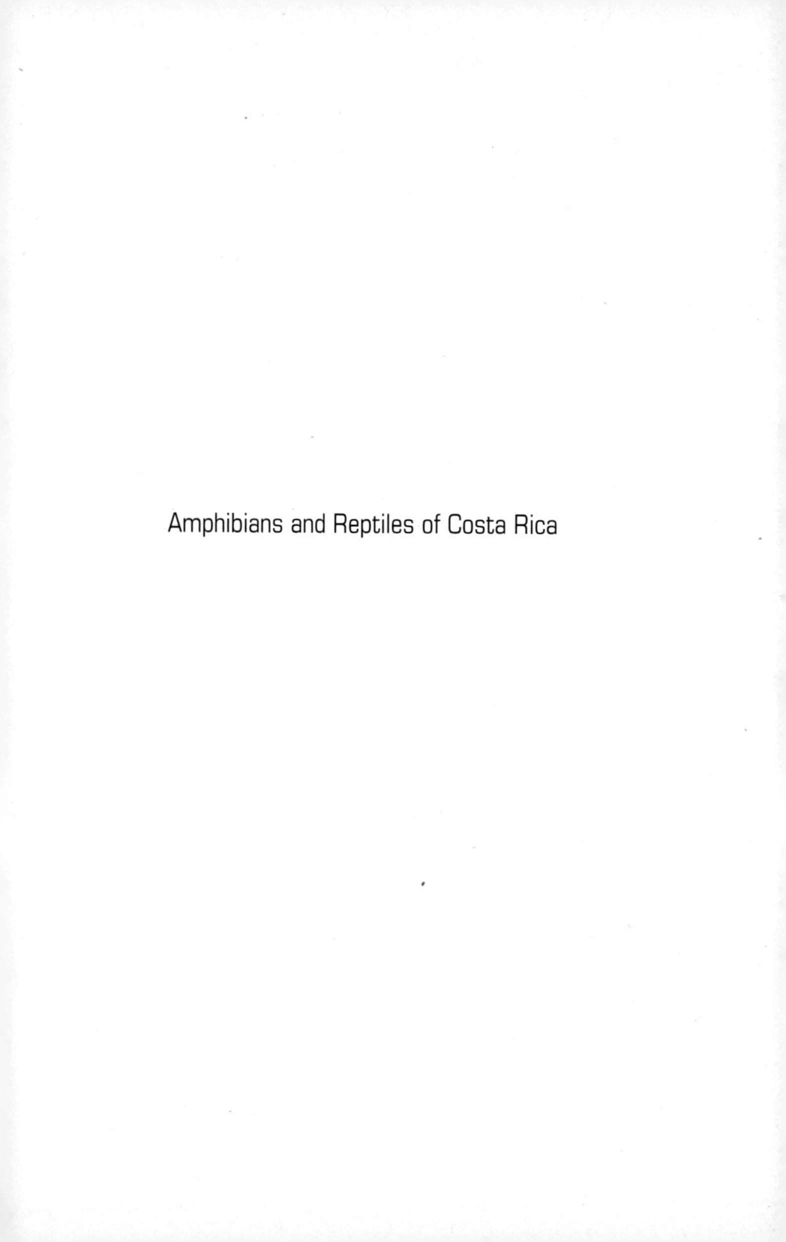

Amphibians and Reptiles of Costa Rica

Amphibians and Reptiles of Costa Rica

A POCKET GUIDE

Federico Muñoz Chacón
Richard Dennis Johnston

A Zona Tropical Publication
from
Comstock Publishing Associates
a division of
CORNELL UNIVERSITY PRESS
Ithaca and London

First published 2013 by Cornell University Press

First printing, Cornell Paperbacks, 2013
Printed in China

Library of Congress Cataloging-in-Publication Data

Muñoz Chacón, Federico.
Amphibians and reptiles of Costa Rica : a pocket guide / Federico Muñoz Chacón, Richard Dennis Johnston.
p. cm.
"A Zona Tropical publication."
Includes index.
ISBN 978-0-8014-7869-7 (pbk. : alk. paper)
1. Amphibians–Costa Rica–Identification. 2. Reptiles–Costa Rica–Identification. I. Johnston, Richard Dennis. II. Title.

QL656.C783M86 2013
597.9097286–dc23

2012039623

Paperback printing 10 9 8 7 6 5 4 3 2 1

Book design: Zona Creativa S.A.
Designer: Gabriela Wattson

Contents

Introduction vii
Notes on Measurements and Symbols xi
About Range Maps xii
Map of Costa Rica xiii

Amphibians 1
- Caecilians 2
- Salamanders 4
- Frogs & Toads 14
  - Burrowing Toads 14
  - Toads 14
  - Dink Frogs 21
  - Rain Frogs 23
  - Robber Frogs 30
  - Foam Frogs 33
  - Dwarf Frogs 35
  - Marsupial Frogs 36
  - Treefrogs 36
  - Glass Frogs 53
  - Poison Frogs 58
  - Rocket Frogs 61
  - Narrow-mouthed Toads 62
  - Pond Frogs 63

Reptiles 66
- Turtles 67
  - Hard Shelled Sea Turtles 67
  - Leatherback Sea Turtle 69
  - Snapping Turtles 70
  - Mud Turtles 71
  - Sliders 73
  - Wood Turtles 74

Lizards 76
Basilisk Lizards 76
Iguanas 79
Spiny Lizards 81
Anolis 82
Canopy Lizards 91
Gecko 92
Night Lizards 96
Skinks 97
Teiids 98
Microteiids 101
Alligator Lizards 103
Snakes 106
Wormsnakes 106
Slender Blindsnakes 107
Blindsnakes 107
Burrowing Python 108
Boas 109
Dwarf Boas 111
Colubrid Snakes 112
Pitvipers 148
Coral Snakes & Sea Snakes 157
Crocodilians 160
Caimans 160
Crocodiles 160

Glossary 161
Acknowledgments 163
Systematic Index 165
About the Authors 172

Introduction

With more than 400 species of amphibians and reptiles, Costa Rica possesses some of the richest herpetological biodiversity on the planet. Several factors combine to make possible such biological wealth. For one, the country has many distinct habitats within varied terrain. In lowlands of the eastern Caribbean and southern Pacific, for example, tropical wet forests receive rain almost year-round, while the tropical dry forest in the Pacific northwest experiences many rainless days. In cloud forests of mountainous regions, in which populations of one species may have lived for eons in isolation from populations of a once-similar species on a neighboring ridge, endemism is relatively common.

Costa Rica is a young country in geological terms, which is another factor in its biological diversity. Some 3 million years ago, the land that today comprises Central America had emerged from the ocean, linking the once-separated continents of North America and South America. Even before this land bridge was entirely in place, animals from the north had begun island hopping south, and those in the south began radiating north, with the result that Costa Rica is today home to many species that originated elsewhere.

Compared with most countries, Costa Rica has done an admirable job of protecting its bounteous natural resources. Well over 20% of its land has been set aside for national parks, biological reserves, and wildlife refuges. In addition, a virtual league of nongovernmental organizations and other private initiatives play a vital role in conservation. And, with ecotourism such an important component of the economy, protecting nature is generally understood to be in the national interest.

Costa Rica's dedication to conservation is just one of many factors that make it a great place to see wildlife. Visitors can take comfort in the fact that this country—one of the few in the world with no army—is a peaceful place with a longstanding democracy and friendly people. And, because Costa Rica is a tiny country with incredible natural diversity, tourists are able to see an amazing collection of animals during even a brief visit. With two coasts—one facing the Caribbean, the other the Pacific—and nesting sites that receive five species of sea turtle, Costa Rica is arguably the best place in the world to see those magnificent creatures. While it is true that many species of amphibians and reptiles are furtive nocturnal creatures that are thus often difficult to see, the country has some of the best nature guides in the world who are adept at helping people see a broad range of species.

Because Costa Rica has many species of amphibians and reptiles—and because so many of these resemble one another—identification can be a challenge. In this book, small and light enough to carry into the field, the authors describe all 418 species found within the borders of Costa Rica. Biologists have proposed adding several new species, and some of these will undoubtedly appear on future lists. Photographs or illustrations are provided for about 80% of the species in the book; those without illustrations are either rare (or for some other reason difficult to see) or, in some cases, extinct. With this book in hand, you stand a much better chance of being able to recognize what you see.

Every effort has been made to provide up-to-date scientific names, a daunting task at a time when phylogenetics continues to bring about many changes in herpetological systematics. More than half of the frogs found within the borders of Costa Rica are now listed under new genera or even placed into new families, and we can expect many more changes in the future. To offer one example, the family Colubridae, which currently contains the greatest number of snake species, will likely be divided into multiple families. In cases for which a species has recently been given a new scientific name, the old scientific name is presented in parentheses.

Unlike scientific names, the common name for a given species can vary from place to place; and, to make matters even more confusing, sometimes the same common name is given to different species. In Costa Rica, for example, the common name lora is ascribed to many green arboreal snakes. To make things less confusing, we have taken into account historical usage and local customs to decide on the best possible common name for a given species.

More than 15% of the snakes in Costa Rica are venomous, and many of these are potentially lethal. Unfortunately, it's not always easy to distinguish venomous snakes from the nonvenomous species, several of which mimic the coloration of coral snakes. And the False Fer-de-Lance, as its name would suggest, is a pitviper mimic. Given the potentially serious consequences of a snake bite, capture or handling of Costa Rican snakes should not be attempted except by people with experience in doing so. Front-fanged venomous species are not the only ones that pose a threat, as there are numerous rear-fanged snakes as well, with enlarged, grooved teeth in the rear of the upper jaw. The venom flows down these teeth and is then chewed into the wound. Although none of the rear-fanged species from Costa Rica are considered dangerous, many can produce unpleasant reactions, including pain, swelling, discoloration, and excessive bleeding.

Many amphibians have defenses against predators in the form of toxic skin secretions. The most notable are the poison frogs, which produce powerful toxins (metabolized from chemicals in their arthropod diet) and secrete them through their skin. Poison frogs found in Costa Rica are less toxic than the species used by some tribes in Panama and South America for blow gun darts, but their secretions can nonetheless prove hazardous if they enter the body, either through a cut or the mouth or eyes.

Although Costa Rica is to be commended for its efforts to conserve and protect natural areas, such activities have not spared the country from the amphibian decline now occurring worldwide. One of the first species to disappear in Costa Rica was the Golden Toad, found only in Monteverde and last seen in the late 1980s. Since then, at least five additional species are thought to be extinct, and many others are considered threatened. There are several causes for this decline, including climate change, habitat loss, overuse of agricultural chemicals, and introduced pathogens. The moral duty of all visitors to natural areas is to respect and protect local ecosystems and to help communicate to others the need to preserve these important animals.

Notes on Measurements and Symbols

All size measurements indicate the distance from an adult's snout to the tip of its tail, except in the case of salamanders and lizards, which are measured from snout to vent.

The jar symbol is used to indicate when a photograph depicts a preserved specimen.

Not all snakes have fangs, but snakes that do have one of three kinds:

Rear fang

Viper fang

Elapid fang

The skull-and-crossbones symbol indicates snakes that are highly venomous.

The following three symbols are also used in this book:

Male

Female

Juvenile

About Range Maps

Range maps are included for every species in this book. Each map is divided into the following seven regions:

Northwest (NW)
Northern Mountains (NM)
Northeast (NE)
Central Mountains (CM)
Southwest (SW)
Southern Mountains (SM)
Southeast (SE)

Regions may be color coded to indicate the elevations at which some species occur:

- Low elevations
- High elevations
- Multiple elevations

Note that range maps often indicate that a species occurs in more than one region, sometimes using more than one color.

The elevation within a given region often varies greatly, as the map on the opposite page clearly shows. And while shading in an entire region seems to suggest that a given species occurs throughout that region, it might actually occur in only a few locations.

Map of Costa Rica

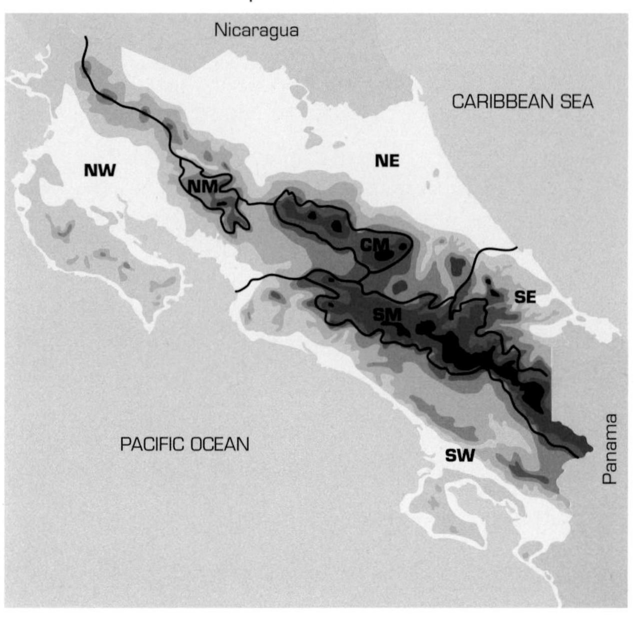

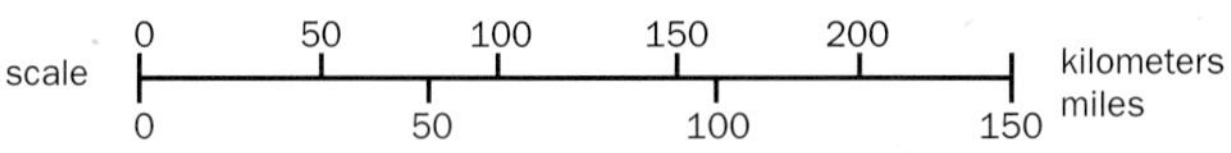

amphibia

Amphibians

191 species

GYMNOPHIONA (7 species) Caecilians

Caeciliidae (7 species) Caecilians

Blind, fossorial, and limbless, all members of the order look like giant earthworms. The small eyes are capable of sensing only light and dark. Glands in the skin produce mucus that contains toxins to ward off predators. These burrowing amphibians rarely surface but are occasionally seen at night during rains. Caecilians feed mainly on earthworms. As caecilians occur only in tropical regions, many people from other parts of the world do not know about them. All species in Costa Rica (known in Spanish as *solda con solda*) belong to the family Caeciliidae; its members have both primary annuli (around the whole body) and secondary annuli (incomplete), as well as a tentacle between each eye and nostril (or closer to either the eye or the nostril). Though some members of the family lay eggs, most (perhaps all) species in Costa Rica are viviparous.

Dermophis costaricense — Costa Rican Caecilian

RDJ

38 cm (15 in)

Eye is visible; each tentacle lies halfway between eye and nostril. Dorsum is gray. Has 107 to 117 primary annuli, 74 to 96 secondary annuli.

Dermophis glandulosus — Southwestern Caecilian

Eye is visible; each tentacle lies halfway between eye and nostril. Dorsum is gray. Has 91 to 106 primary annuli, 37 to 60 secondary annuli.

40 cm (16 in)

Dermophis gracilior — Southern Caecilian

Eye is visible; each tentacle lies halfway between eye and nostril. Dorsum is gray. Has 91 to 117 primary annuli, 65 to 78 secondary annuli.

34 cm (14 in)

Dermophis occidentalis — **Western Caecilian**

25 cm (10 in)

RDJ

Eye is visible; each tentacle lies halfway between eye and nostril. Dorsum is gray; head is paler than rest of body. Has 95 to 112 primary annuli, 29 to 37 secondary annuli.

Dermophis parviceps — **Small-headed Caecilian**

22 cm (9 in)

TL

Eye is visible; each tentacle lies halfway between eye and nostril. Dorsum is purplish; head is pink. Has 85 to 102 primary annuli, fewer than 27 secondary annuli.

Gymnopis multiplicata — **Purple Caecilian**

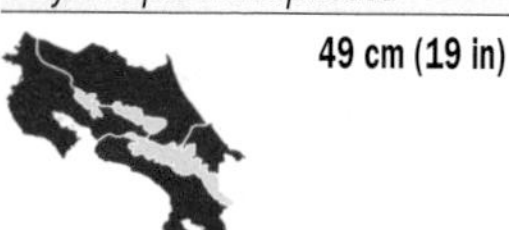

49 cm (19 in)

JPM

Eye not visible; tentacles are posterior to nostril. Dorsum is purplish; venter varies from pink to cream. Has 112 to 133 primary annuli, 84 to 117 secondary annuli.

Oscaecilia osae — **Osa Caecilian**

38 cm (15 in)

JCS

Eye not visible; each tentacle is located below the nostril. Purplish overall, with lighter venter and anterior region; head is pink. Has 232 primary annuli; secondary annuli sometimes present. Occurs only on the Osa Peninsula.

Plethodontidae (43 species)
Lungless Salamanders

Salamanders are descendants of the vertebrates that first colonized land. They resemble lizards but have moist smooth skin instead of scales. While most species occur in temperate zones, many members of the family Plethodontidae occur in tropical regions of the Americas. Species in this family are lungless and respire through the skin, a feature found outside of the family Plethodontidae in only a few other salamander species. Another characteristic feature is a groove (naso-labial groove) that runs from each nostril toward the lip. These salamanders are oviparous (development occurs in the egg, which is laid in moist substrate). Some species in this family are terrestrial while others are arboreal. These secretive animals are generally small and slender, with short legs; most species are dark (gray, brown, and black tones predominate). All species in Costa Rica belong to this family. Species in the genus *Bolitoglossa* have fewer than 16 lateral grooves and have broad hands and feet; those in the genus *Nototriton* also have fewer than 16 lateral grooves but have narrow hands and feet; and species in the genus *Oedipina* have 16 or more lateral grooves. The majority of species in Costa Rica are known only from a characteristic habitat (or occur in vicinity of same).

Bolitoglossa alvaradoi — Alvarado's Salamander

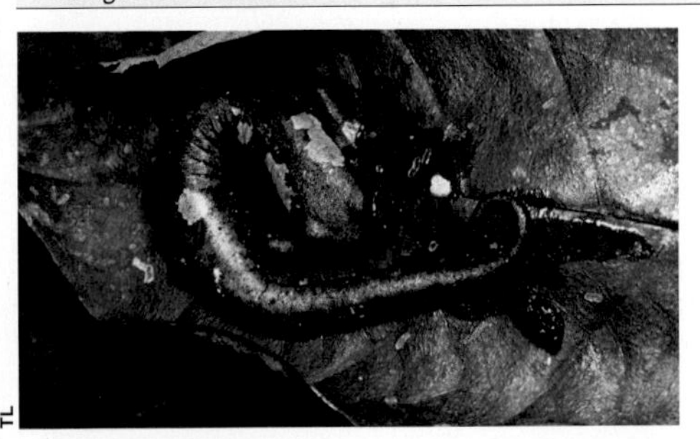

TL

7.5 cm (3 in)

Shows some light blotches on dark dorsum. Venter dark. Has complete interdigital webbing. Nocturnal. Occurs on the Caribbean slope, from Tilarán to Moravia de Chirripó. This arboreal species is sometimes found in bromeliads.

Bolitoglossa bramei — Brame's Salamander

Uniformly yellowish brown, with or without fine light lines along body. Occurs on both sides of CR–Panama border, in a small area between Cerro Pando and Volcán Barú.

4 cm (1.6 in)

Bolitoglossa cerroensis — Mountain Salamander

7 cm (2.8 in)

AV

Purple-gray body color, with irregular light lines on sides. Occurs in Cerro de la Muerte area, at about 9,800 ft. Often found at base of rocky cliffs.

Bolitoglossa colonnea — Eyebrowed Salamander

5 cm (2 in)

AV

Uniformly brownish gray. Has complete interdigital webbing. Shows a characteristic fleshy fold on the front of the head. Often found close to the ground on vegetation.

Bolitoglossa compacta — Compact Salamander

7 cm (2.8 in)

FMC (RVL)

Dorsum and venter are brown; irregular rusty blotches (of varying sizes) appear along length of body; has red or orange tip on the tail. Occurs on both sides of CR–Panama border, from Cerro Pando to Valle del Silencio. Found mostly under logs.

Bolitoglossa diminuta — Tiny Salamander

3 cm (1.2 in)

This is the smallest salamander in the genus *Bolitoglossa*. Dorsum is brown; a broad darker band runs along each side of the body. Found only in Tapantí (Cartago), at 4,900 ft. Apparently dwells in forest canopy.

Bolitoglossa epimela — Black-backed Salamander

4.5 cm (1.8 in)

Shows uniform coloration that varies, individual to individual, from dark brown to black. Has complete interdigital webbing. Known only from area between Turrialba and Tapantí (Cartago); found at 2,950 ft. Nocturnal. Forages on logs or close to ground on vegetation; extensive webbing on digits allows it to walk on the underside of leaves.

Bolitoglossa gomezi — Gomez's Salamander

5 cm (2 in)

Reddish brown, with or without fine light lines along body. Occurs on both sides of CR–Panama border, in a small area between Las Cruces and Río Candela.

Bolitoglossa gracilis — Slender Salamander

4 cm (1.6 in)

Brown coloration turns yellowish toward dorsum. Occurs in Tapantí (Cartago), at 4,200 ft. Found in mossy environments.

Bolitoglossa lignicolor — Wood Colored Salamander

TL

7.5 cm (3 in)

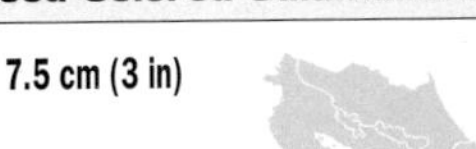

Brown overall with light-brown dorsal blotches. Occurs in lowlands of the Pacific slope, in central and southern regions. Mostly found on trees, either within bromeliads or near them.

Bolitoglossa marmorea — Marbled Salamander

7 cm (2.8 in)

Light dorsum and venter, with dark marble streaking on flanks. Occurs on Cerro Pando, along the Panamanian border. Nocturnal. At night, found close to ground on vegetation; during the day, found under logs and rocks.

Bolitoglossa minutula **Miniature Salamander**

3.5 cm (1.4 in)

FMC

Most individuals are uniformly dark; some individuals show reddish dorsal area. Has complete interdigital webbing. Occurs on Cerro Pando, along the Panamanian border. Nocturnal. Found close to ground hiding within vegetation.

Bolitoglossa nigrescens **Black Salamander**

9 cm (3.5 in)

AV

Uniformly black, showing no spots or blotches. Occurs in Talamanca Range, at 6,600 ft. Terrestrial.

Bolitoglossa obscura **Tapantí Giant Salamander**

9 cm (3.5 in)

Large, uniformly black salamander. Has complete interdigital webbing. Known from single specimen collected near Tapantí at 5,100 ft.

Bolitoglossa pesrubra **Brown Salamander**

6.5 cm (2.6 in)

FMC

oloration varies from dark brown to ght brown, with some individuals so showing blotches; legs and feet tend to be pink (to red). Found along Talamanca Range; common ound dweller on Cerro de la Muerte. Very similar to *B. subpalmata* (p. 9).

Bolitoglossa robusta — Ring-tailed Salamander

AV

13 cm (5 in)

A large salamander. Uniformly black except for light ring around base of tail. Occurs from 4,900 to 6,600 ft in Guanacaste, Central, and Talamanca mountain ranges. Nocturnal. Found on ground or on low vegetation.

Bolitoglossa schizodactyla — Wake's Salamander

FMC

6 cm (2.4 in)

Brown dorsum shows light yellow blotches; venter yellow.

Bolitoglossa sombra — Shadowy Salamander

Large, uniformly black salamander, with extensive webbing on feet. Occurs in eastern regions of Talamanca Range, at 6,600 ft.

7.5 cm (3 in)

Bolitoglossa sooyorum — Yellow-dotted Salamander

AV

6.5 cm (2.6 in)

Purplish brown overall, with many yellow dots and blotches. Large feet. Occurs only at high elevations (about 9,800 ft) of Talamanca Range.

Bolitoglossa striatula **Striated Salamander**

6.5 cm (2.6 in)

DH

Yellowish brown overall, with fine dark-brown longitudinal stripes along the body. Has complete interdigital webbing. Forages on low vegetation. If threatened, turns upside down and falls from vegetation, perhaps as a means of escaping capture.

Bolitoglossa subpalmata **La Palma's Salamander**

6 cm (2.4 in)

FMC

Coloration varies from dark brown to light brown, with some individuals also showing blotches; legs and feet tend to be pink (to red). Occurs in Central and Guanacaste mountain ranges. Very similar to *B. pesrubra* (p. 7) but no range overlap.

Bolitoglossa tica **Tico Salamander**

5 cm (2 in) Variable coloration; individuals range from uniformly brown to brown with red flecks. Occurs between Tapantí and Cerros de Escazú, in northern region of Talamanca Range.

Nototriton abscondens **Concealing Moss Salamander**

3.3 cm (1.3 in) Coloration varies; most individuals show a brown background, some also show fine light longitudinal lines. Occurs in Central Mountain Range, from 3,300 to 8,000 ft.

Nototriton gamezi — **Monteverde Moss Salamander**

2.5 cm (1 in)

Coloration varies from light brown to dark brown, with or without dark spots or blotches. Occurs in Monteverde. Found under moss, from ground level to forest canopy.

Nototriton guanacaste — **Guanacaste Moss Salamander**

3.4 cm (1.3 in)

Relatively robust body. Well-defined head, eyes, legs, and feet. Only 5 individuals known, from Orosí and Cacao volcanoes, between 4,600 and 5,250 ft.

Nototriton major — **Large Moss Salamander**

4 cm (1.6 in)

Known from only one individual found in Talamanca Range, in Quebrada Platanillo (Moravia de Chirripó), somewhere between 3,600 and 4,000 ft.

Nototriton picadoi — **Picado's Moss Salamander**

GSM

3 cm (1.2 in)

Brown overall, with some light lines on head and base of tail. Has been collected from trees, both from moss and from bromeliads. Occurs in Estrella (Cartago), between 5,000 and 6,500 ft.

Nototriton richardi — **Richard's Moss Salamander**

2.4 cm (1 in)

Relatively short trunk; digits of hands and feet connected by membrane; large nostrils. Occurs on Caribbean slope of Central Mountain Range, from 4,000 to 5,600 ft.

Nototriton tapanti **Tapantí Moss Salamander**

2.4 cm (1 in) Has relatively large nostrils. Occurs in Río Orosi Valley (northern Talamanca Range), at 4,200 ft.

Oedipina alfaroi **Alfaro's Worm Salamander**

6 cm (2.4 in) Has 19 to 22 lateral grooves; no maxillary teeth. Occurs in southern Caribbean region.

Oedipina alleni **Allen's Worm Salamander**

5.5 cm (2.2 in)

Note brown and cream streaks on dorsal surface; light markings on head and tail; and 18 lateral grooves.

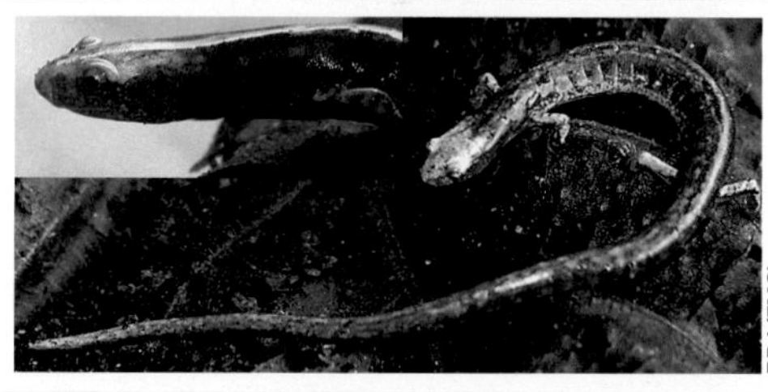

RDJ (FMC)

Oedipina altura **Highland Worm Salamander**

6 cm (2.4 in) Shows a uniform coloration, but color varies (black to brown) from individual to individual. Has 19 to 22 lateral grooves and 25 or more maxillary teeth. Occurs in El Empalme (Cerro de la Muerte, Talamanca Range), at 7,500 ft.

Oedipina carablanca **Los Diamantes Worm Salamander**

6 cm (2.4 in) Dark with white area on head. Has either 17 or 18 lateral grooves. Occurs in northern region of Caribbean slope, in lowland rainforest.

Oedipina collaris **Collared Worm Salamander**

7.6 cm (3 in) Generally uniformly dark, though some individuals show a pale collar. Has 19 to 22 lateral grooves and 80 to 98 maxillary teeth.

Oedipina cyclocauda **Round-tailed Worm Salamander**

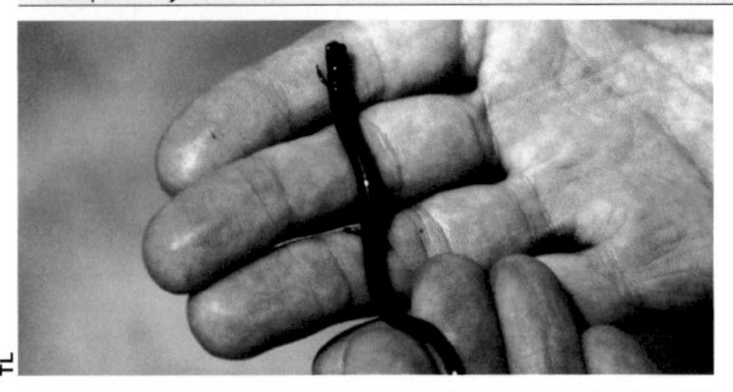

TL

4.5 cm (1.8 in)

Dorsal surface is uniformly dark. Has 19 to 22 lateral grooves and 28 to 55 maxillary teeth.

Oedipina gracilis **Slender Worm Salamander**

4.6 cm (1.8 in)

Shows striated dark coloration. Has 20 lateral grooves and 18 or 19 maxillary teeth. Only a few individuals known, near Guápiles and La Selva.

Oedipina grandis **Large Worm Salamander**

7.1 cm (2.8 in)

Uniformly dark, but with white dots along the length of the body (extent of dots likely variable from individual to individual). Has 19 to 22 lateral grooves. Occurs in Talamanca Range, from 5,900 to 6,500 ft.

Oedipina pacificensis **Pacific Worm Salamander**

AV

5 cm (2 in)

Uniformly dark on dorsal surface, with darker vertebral line along the length of the body. Tail is twice as long as the body. Has 12 to 14 maxillary teeth. Occurs in southern region of Pacific slope.

Oedipina paucidentata **Long Worm Salamander**

6 cm (2.4 in)

Small hind feet; 17 or 18 lateral grooves; 33 to 43 maxillary teeth. Occurs at El Empalme (Talamanca Range), at 7,200 ft.

Oedipina poelzi — **Poelz's Worm Salamander**

6.4 cm (2.5 in)

RVL

Dorsal surface (uniformly dark) is separated from even darker venter by a narrow lateral line. Has 19 to 22 dorsal grooves and 27 to 70 maxillary teeth. Occurs on Caribbean slope, from 4,000 to 7,000 ft.

Oedipina pseudouniformis — **Lowland Worm Salamander**

5.2 cm (2 in)

Uniformly dark on dorsal surface; hind legs relatively long compared with other members of the genus. Has 19 to 22 dorsal grooves and 28 to 55 maxillary teeth. Occurs in lowland rainforest on Caribbean slope.

Oedipina savagei — **Savage's Worm Salamander**

3.8 cm (1.5 in)

Dorsal surface reddish brown, speckled with minute silver dots. Occurs in San Vito de Java region (Puntarenas).

Oedipina uniformis — **Common Worm Salamander**

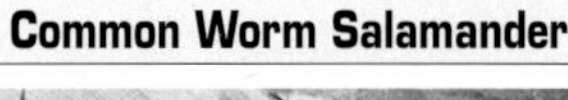

5.7 cm (2.2 in)

AV

Dorsal surface is dark brown; hind legs are very small; has 19 to 22 lateral grooves. Occurs throughout the country, up to 3,300 ft.

Rhinophrynidae (1 species)
Burrowing Toads

This family contains a single primitive species. The burrowing toad is similar in appearance habits to other toads but is distinguished from them by vertical pupils. And, more generally, distinguished from other anurans by a tongue that is attached at the rear of the mouth inst of the front. Fossorial and nocturnal, this toad is usually seen only during the rainy season.

Rhinophrynus dorsalis — **Mexican Burrowing Toad**

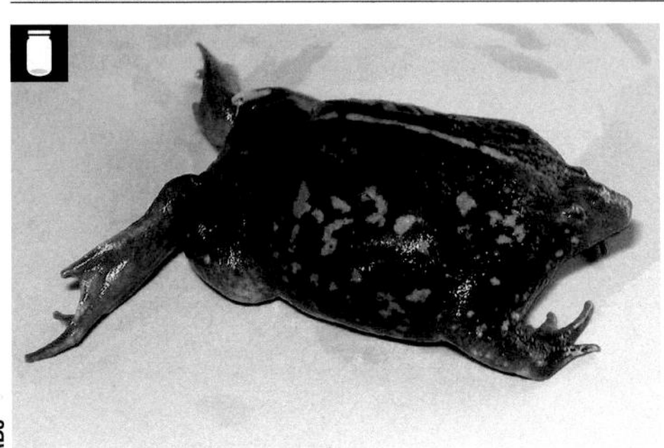
RDJ

8 cm (3.1 in)

With its robust body and loose skin, this toad somewhat resembles a ball from which a little head and 4 tiny limbs project. The head is adapted for probing into spaces that contain its prey: ants and termites. This burrowing species stays underground most of the time, surfacing during the rainy season to mate in temporary ponds.

Bufonidae (18 species)
Toads

Toads generally have rounded bodies, thick warty skin, and large parotid glands. The grea majority of species are nocturnal, although such species are sometimes seen during th day. Unlike most frogs, toads have no teeth on their jaws. They also differ from the majori of frog species in their behavior: toads take short hops, frogs make long jumps. Toads hav horizontal pupils (see, however, burrowing toads, above).

Atelopus chiriquiensis — **Chiriquí Harlequin Frog**

Background coloration is highly variable, from uniform green to yellowish brown, with red and yellow spots or blotches. Small parotid glands; smooth head; lacks tympanum. Diurnal. Found along streams, where it walks (this species does not jump). Possibly extinct.

4.6 cm (1.8 in)

Atelopus chirripoensis — Chirripó Harlequin Frog

4.2 cm (1.7 in)

Dorsum dark brown to black; venter light cream to orange. No tympanum. Known only from near Cerro Chirripó in the Talamanca Range. Diurnal. Found along streams, where it walks (this species does not jump).

Atelopus senex — Green Harlequin Frog

4 cm (1.6 in)

Coloration green, bluish green, or very dark green. Medium-size parotid glands; smooth head; lacks tympanum. Diurnal. Found along streams, where it walks (this species does not jump). Possibly extinct.

Atelopus varius — Yellow & Black Harlequin Frog

5 cm (2 in)

Background coloration is variable—green, yellow, or red—with black blotches. Very small parotid glands; smooth head; lacks tympanum. Diurnal. Found along streams, where it walks (this species does not jump).

TL

Crepidophryne chompipe — Chompipe Pygmy Forest Toad

3.5 cm (1.4 in)

Individuals are uniformly brown or gray. Very small parotid glands; small tympanum. Webbing extends the full length of some (but not all) toes. Occurs in Central Mountain Range.

Crepidophryne epiotica **Talamanca Pygmy Forest Toad**

RDJ

3.5 cm (1.4 in)

Individuals are either uniformly brown or gray. Very small parotid glands; small tympanum. Most toes connected by webbing. Occurs in Talamanca Range.

Crepidophryne guanacaste **Guanacaste Pygmy Forest Toad**

Individuals are uniformly brown or gray. Very small parotid glands; small tympanum. Most toes connected by webbing. Known only from Guanacaste Mountain Range.

3.5 cm (1.4 in)

Incilius (Bufo) aucoinae **Rain Forest Toad**

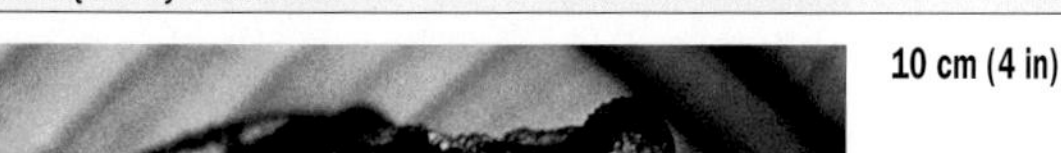

RDJ

10 cm (4 in)

Dorsal coloration varies from yellow to dark brown, sometimes with very dark patches. Parotid glands are small; tympanum present. Cranial crests are low. Has well-defined tubercles on hands and feet.

Incilius (Bufo) coccifer **Dry Forest Toad**

8 cm (3.1 in)

AV

Dorsal coloration varies from yellow to dark brown. Tubercles on hands and feet are high and pointed. Tympanum present. During the rainy season, it reproduces in ponds; during the rest of the year, it remains underground.

Incilius (Bufo) coniferus **Green Climbing Toad**

9.4 cm (3.7 in)

RDJ

Dorsal coloration varies from green to yellowish green. Tubercles on hands and feet are rounded. Lateral tubercles are big, with enlarged spines. Found most often low on vegetation. Feeds on ants.

Incilius (Bufo) fastidiosus **Talamanca Toad**

5.5 cm (2.2 in)

Black background coloration; tips of parotid glands and tubercles are reddish brown. Cranial crest high and massive; pale venter is mottled with black. Tubercles are not well developed. Possibly extinct.

Incilius (Bufo) holdridgei **Holdridge's Toad**

RVL

5 cm (2 in)

Dorsum has black background; tips of parotid glands and tubercles are reddish brown; venter is pale; black blotches on chest. Low cranial crests. No tympanum. Has poorly developed tubercles. Known only from Barva Volcano.

Incilius (Bufo) luetkenii **Yellow Toad**

RDJ

RDJ

10 cm (4 in)

Dorsal coloration varies from yellow to dark brown, sometimes with very dark patches. Parotid glands are small. Tympanum present. Tubercles are high and pointed.

Incilius (Bufo) melanochlorus — Wet Forest Toad

10 cm (4 in)

Dorsal coloration varies from yellow to dark brown, sometimes showing dark areas. Parotid glands are small. Tympanum present. Tubercles are high and pointed.

Incilius (Bufo) periglenes — Golden Toad

6 cm (2.4 in)

EXTINCT

Known only from Monteverde.

Incilius (Bufo) valliceps **Gulf Coast Toad**

AV

7.6 cm (3 in)

Dorsal background coloration highly variable (uniform reddish, brown, or gray); most individuals have a light vertebral stripe; venter light overall with black spots. Small parotid gland; cranial crests present. Tubercles are well defined.

Rhaebo (Bufo) haematiticus **Litter Toad**

RDJ

7.6 cm (3 in)

Brown coloration overall; a black band covers sides of face, giving it the appearance of a dry leaf. Very large parotid gland. Lacks cranial crests and tubercles.

Rhinella (Bufo) marina **Marine Toad**

AV (FMC)

18 cm (7.1 in)

Uniformly brown. Parotid glands large; tympanum present. Tubercles are well defined and pointed. The largest and most common toad in Costa Rica. This nocturnal toad is especially abundant near human habitations, where it has learned to approach lights in order to catch insects.

Eleutherodactylidae (6 species)
Dink Frogs

Dink frogs have long slender fingers with no webbing. They undergo direct development, with no free tadpole stage. Eggs are deposited in moist substrates. Costa Rican species in this family are known for their small size and very loud calls. As they are largely arboreal (and nocturnal), they are more often heard than seen. Pupils are horizontal.

Diasporus (Eleutherodactylus) diastema — **Common Dink Frog**

2.5 cm (1 in)

Uniform coloration varies from cream to brown, depending on individual. Except for expanded, rounded finger disks, has no other prominent features. Makes loud *dink* call from perches low on vegetation. Abundant.

DH

Diasporus (Eleutherodactylus) hylaeformis — **Montane Dink Frog**

2.5 cm (1 in)

FMC

Color ranges from grayish yellow to yellowish red. Smooth skin. Similar to *D. diastema* (above), but finger disks triangular, not round. Makes loud *dink* call.

Diasporus (Eleutherodactylus) tigrillo — **Margay Dink Frog**

2 cm (0.8 in) Yellow-orange coloration overall, with tiny dark spots. Smooth skin. Makes loud *dink* call.

Diasporus ventrimaculatus **Mottled-bellied Dink Frog**

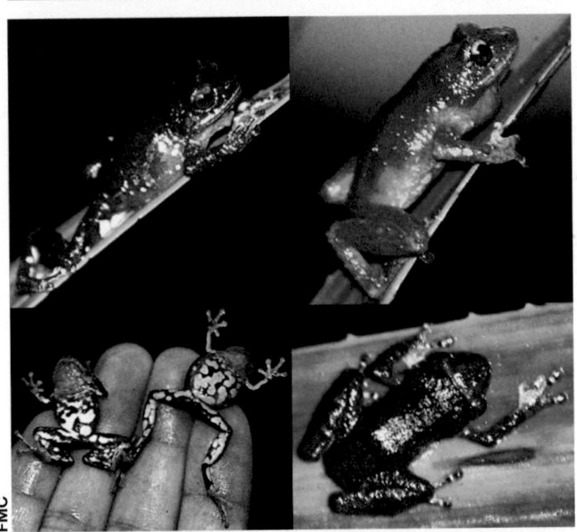

FMC

2.5 cm (1 in)

Color is variable; individuals can be white, red, black, or other colors. Makes loud *dink* call. Known only from Valle del Silencio.

Diasporus (Eleutherodactylus) vocator **Vocal Dink Frog**

AV

2 cm (0.8 in)

Dorsal pattern is variable; some individuals are uniformly dark, others show gray or brown blotches. Makes loud *dink* call.

Eleutherodactylus coqui **Coqui Frog**

INTRODUCED

3.5 cm (1.4 in)

Large for a dink frog. Color ranges from light to medium shades of gray (or brown). Most individuals are uniform in color; some individuals show some kind of pattern and/or a light middorsal stripe. Terrestrial and arboreal. Known for its loud *KoKee* call. Introduced near Turrialba.

Craugastoridae (28 species)
Rain Frogs

Rain frogs have long slender fingers with no webbing. Pupils are horizontal. Frogs in this family vary a great deal in size and appearance. They undergo direct development, with no free tadpole stage. Eggs are deposited in moist substrates. Rain frogs tend to be nocturnal but individuals are sometimes seen during the day. Costa Rican species are largely terrestrial.

Craugastor (Eleutherodactylus) andi — **Salmon-bellied Rain Frog**

8 cm (3.1 in) Shows bright yellow spots or stripes on posterior thigh. The 2 outer finger disks of each hand are much larger than the 2 inner finger disks. Arboreal.

Craugastor (Eleutherodactylus) angelicus — **Tilarán Rain Frog**

7.5 cm (3 in) Ventral coloration often yellow to orange. All finger disks are wider than the digits.

Craugastor (Eleutherodactylus) bransfordii — **Bransford's Litter Frog**

2.5 cm (1 in)

FMC

Very small. Dark brown overall, with some individuals showing a middorsal stripe. Note obvious tubercles projecting from hands and feet. Similar to *C. polyptychus* (p. 28).

Craugastor (Eleutherodactylus) catalinae — **Lips' Rain Frog**

7.5 cm (3 in) Venter light in color. Also note webbing on toes and large finger disks.

Craugastor (Eleutherodactylus) crassidigitus — **Slim-fingered Rain Frog**

RDJ

RDJ

5 cm (2 in)

Coloration highly variable, although in most cases posterior thighs show a rust color. Also note webbing on toes. Similar in appearance to a desiccated leaf, this species spends most of its time in leaf litter on the forest floor, where it is common.

Craugastor (Eleutherodactylus) cuaquero — **Monteverde Rain Frog**

Thighs have many small yellow dots. Light stripe on throat. Long legs.

5 cm (2 in)

Craugastor (Eleutherodactylus) escoces — **Scott's Rain Frog**

EXTINCT

Venter red. Finger disks large. Some webbing between toes.

6.5 cm (2.5 in)

Craugastor (Eleutherodactylus) fitzingeri — **Common Rain Frog**

FMC

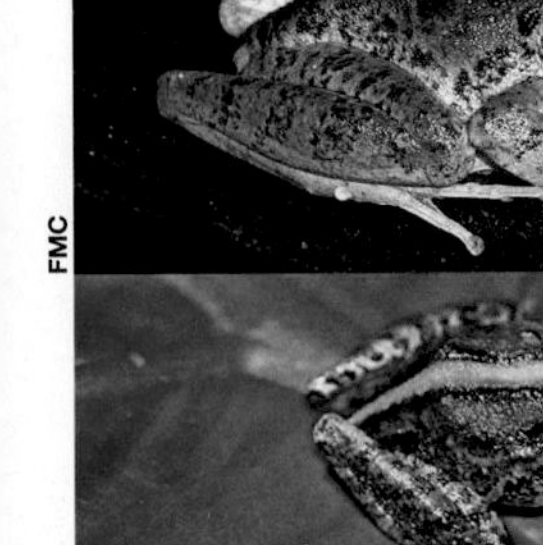

RDJ

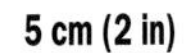

5 cm (2 in)

Rugose dorsum. Coloration highly variable, though in most cases posterior thighs are brown with light spots. Some individuals have a middorsal stripe. Note pale throat stripe. Moderate toe webbing.

Craugastor (Eleutherodactylus) fleischmanni **Fleischmann's Rain Frog**

7 cm (2.8 in) Venter light yellow. Some webbing on toes.

Craugastor (Eleutherodactylus) gollmeri **Gollmer's Rain Frog**

5 cm (2 in)

FMC

Note dark eye mask that extends to each axilla; upper half of the iris is red. Long legs. Diurnal.

Craugastor (Eleutherodactylus) gulosus **Giant Rain Frog**

10 cm (4 in) Broad head; dark coloration; lacks webbing on toes. Similar to *C. megacephalus* (below).

Craugastor (Eleutherodactylus) megacephalus (biporcatus) **Broad-headed Rain Frog**

6.3 cm (2.5 in)

RDJ (RDJ)

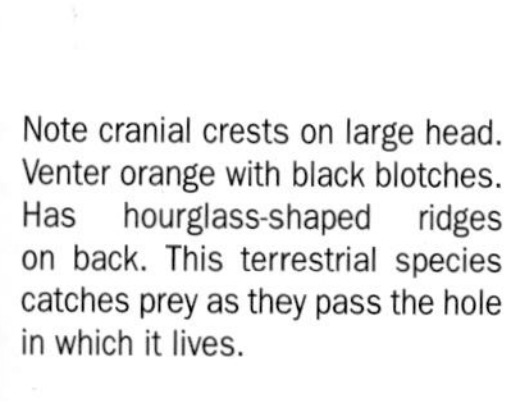

Note cranial crests on large head. Venter orange with black blotches. Has hourglass-shaped ridges on back. This terrestrial species catches prey as they pass the hole in which it lives.

Craugastor (Eleutherodactylus) melanostictus **Quark Frog**

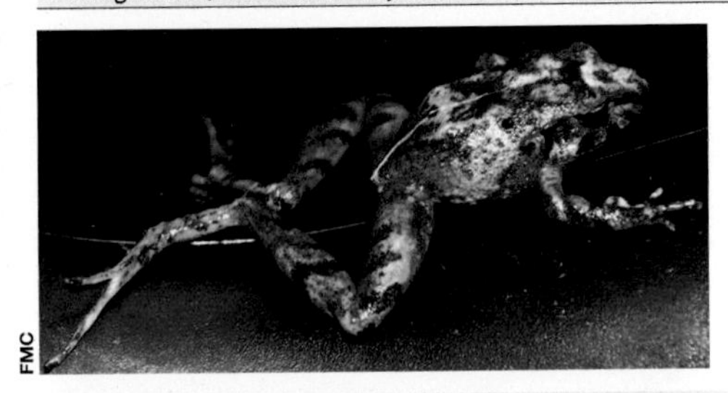

FMC

5 cm (2 in)

Note many small black dots on back; very prominent black bars on thighs extend to posterior surface.

Craugastor (Eleutherodactylus) mimus **Mimic Rain Frog**

AV

6 cm (2.4 in)

Pointed head; black eye mask extends almost to groin. Similar to *C. noblei* (below).

Craugastor (Eleutherodactylus) noblei **Noble's Rain Frog**

TL

6.5 cm (2.6 in)

Pointed head; golden iris; and black eye mask that reaches beyond axilla.

Craugastor (Eleutherodactylus) obesus (punctariolus) **Obese Rain Frog**

8.6 cm (3.4 in) Rugose dorsal skin is brown with dark spots; smooth venter is yellow. Also note large finger disks and webbing on toes.

Craugastor (Eleutherodactylus) persimilis **Little Brown Rain Frog**

2 cm (0.8 in)

TL

Dark-brown dorsal skin shows numerous tubercles; thighs mottled; groin red. Second finger is longer than the first.

Craugastor (Eleutherodactylus) phasma **Phantom Rain Frog**

5 cm (2 in) White body and black eyes, known from only one specimen.

Craugastor (Eleutherodactylus) podiciferus **Piglet Litter Frog**

3.8 cm (1.5 in)

RDJ

Robust body. Brown (to light brown) dorsal pattern shows some dark blotching.

Craugastor (Eleutherodactylus) polyptychus — **Rain Frog**

TL

2.5 cm (1 in)

Very small. Brown overall but note red groin. Has large tubercles that project from hands and feet. This diurnal species occurs in leaf litter. Similar to *C. bransfordii* (p. 23).

Craugastor (Eleutherodactylus) ranoides (rugulosus) — **Lowland Rain Frog**

RP

RP

7.5 cm (3 in)

Has a blunt snout. Posterior section of thighs is brown with light spots. Venter yellow.

Craugastor (Eleutherodactylus) rayo — **Lightning Rain Frog**

Dorsal surface is smooth and brown. Thighs purplish. Also note light middorsal stripe.

6.4 cm (2.5 in)

Craugastor (Eleutherodactylus) rhyacobatrachus — **Torrent Rain Frog**

Brown dorsum; venter pale yellow with brown markings; posterior of thighs mottled with yellow and brown.

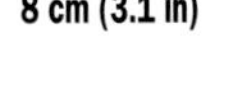

8 cm (3.1 in)

Craugastor (Eleutherodactylus) rugosus **Rugose Rain Frog**

6.8 cm (2.7 in)

AV (AV)

Dorsum brown and extremely rugose. Thighs show black and red stripes.

Craugastor (Eleutherodactylus) stejnegerianus **Stejneger's Rain Frog**

2.5 cm (1 in)

FMC

Rugose dorsal skin varies in color, with light- to dark-brown blotches. Thighs have contrasting brown and light brown bars. Groin is reddish.

Craugastor (Eleutherodactylus) talamancae **Talamanca Rain Frog**

5 cm (2 in)

RDJ (RDJ)

Dorsal coloration brown or light brown depending on individual; venter white to yellow. Also note elongate body, long legs, and pointed head.

Craugastor (Eleutherodactylus) taurus — **Bull Rain Frog**

TL

7.8 cm (3.1 in)

Skin is rugose. Dorsal pattern gray to brown with dark blotches; pale venter is smooth; thighs show blotches.

Craugastor (Eleutherodactylus) underwoodi — **Underwood's Rain Frog**

FMC

3 cm (1.2 in)

Rugose dorsal surface is variable in coloration, showing light- to dark-brown blotches. Thighs with contrasting brown and pale brown bars. Groin reddish.

Strabomantidae (9 species)
Robber Frogs

The long slender fingers of robber frogs lack webbing. Undergoes direct development, with no free tadpole stage. Eggs are deposited in moist substrates. Pupils are horizontal. Several species resemble a treefrog, one mimics a poison frog, and yet another has a toadlike appearance. Mostly active at night but sometimes seen during the day. Generally seen on the ground but sometimes found low on vegetation.

Pristimantis (Eleutherodactylus) altae — **Coral-spotted Robber Frog**

Dark coloration overall, with yellow (to red) patch on the groin.

2.5 cm (1 in)

Pristimantis (Eleutherodactylus) caryophyllaceus **Leaf-breeding Robber Frog**

2.5 cm (1 in)

AV

Most individuals are light brown, but coloration is highly variable. Long snout. Also note long tubercle on each heel.

Pristimantis (Eleutherodactylus) cerasinus **Horn-nosed Robber Frog**

3.8 cm (1.5 in)

AV

Groin is red. Has long legs and a long tubercle on each heel.

Pristimantis (Eleutherodactylus) cruentus **Golden-groined Robber Frog**

3.8 cm (1.5 in)

FMC

DH

Coloration highly variable. Groin and thighs have gold and brown speckling. Has a large tubercle on each heel; also note tubercle above each eye.

Pristimantis (Eleutherodactylus) gaigei — False Poison Frog

RDJ

3.8 cm (1.5 in)

An orange-red stripe runs along each side of dark-brown back. An apparent mimic of similar-looking frogs in the genus *Phyllobates* (pp. 59-60).

Pristimantis (Eleutherodactylus) moro — Red-headed Robber Frog

Green overall, with red on head and on distal half of legs. Resembles a glass frog (p. 53).

2.5 cm (1 in)

Pristimantis (Eleutherodactylus) pardalis — Spotted Robber Frog

Dark gray to black frogs with large white spots on groin and thighs.

2.5 cm (1 in)

Pristimantis (Eleutherodactylus) ridens — Pygmy Robber Frog

RDJ

2.5 cm (1 in)

A yellow-green frog with contrasting dark-brown blotch on each tympanic membrane. Has a very pointed nose.

Strabomantis (Eleutherodactylus) bufoniformis — Toad-like Robber Frog

Entire dorsum is covered with white-tipped tubercles; body has a very granular look. Some webbing between toes.

9 cm (3.5 in)

Leptodactylidae (5 species)
Foam Frogs

Foam frogs have long slender fingers that lack webbing. Pupils are horizontal. Foam frogs have minimal (or no) toe webbing but are otherwise similar in appearance to frogs in the family Ranidae (p. 63). Generally found on the ground. Nocturnal but on rare occasions can be seen during the day. Females in the genus *Leptodactylus* lay their eggs in foamy structures on wet ground or in water.

Leptodactylus bolivianus — Marbled Foam Frog

10 cm (4 in)

AV

Note large dark spots on yellowish back and thighs. Yellow upper lip. Smooth skin.

Leptodactylus fragilis (labialis) — White-lipped Frog

3.8 cm (1.5 in)

AV

RDJ (RDJ)

Coloration and pattern are variable. Note pointed head and distinctive white upper lip; also note light stripe on posterior of thigh. Has rows of tubercles on blotchy back.

Leptodactylus melanonotus **Black-backed Frog**

RDJ

5 cm (2 in)

Dark-brown frog overall; dorsum shows even darker blotches and some tubercles. Iris is golden above and brownish below.

Leptodactylus poecilochilus **Brown Foam Frog**

AV

5 cm (2 in)

Overall color ranges from gray to brown; some individuals show a light middorsal stripe. One to 3 dorsolateral folds.

Leptodactylus savageii (pentadactylus) **Savage's Bull Frog**

18 cm (7.1 in)

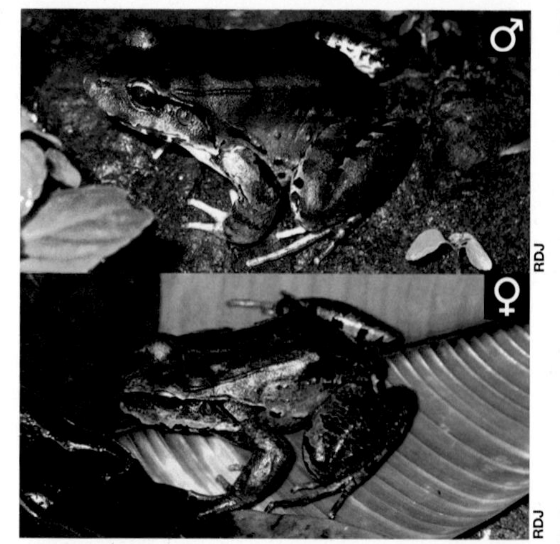

Smooth dorsum has dark blotches on a yellowish-green background. Also note dorsolateral ridges. Males have a red spot on each side of the body and spines on the chest. Diet includes other frogs, small birds, and even mammals.

Leiuperidae (1 species)
Dwarf Frogs

Dwarf frogs are small toadlike frogs with a series of glands along the sides of the body. Other distinguishing features include horizontal pupils, absence of finger disks, and no webbing on fingers or toes. Nocturnal. Females lay eggs in foamy structures on water.

Engystomops (Physalaemus) pustulosus **Tungara Frog**

3.3 cm (1.3 in)

Rugose dorsal surface is blotchy brown. Also note distinct tarsal tubercle. Females deposit eggs in foamy structures on or near water.

Hemiphractidae (1 species)
Marsupial Frogs

On marsupial frogs, the head resembles a crown. Has webbing on toes but lacks webbing on fingers. Pupils are horizontal. Females of this family carry their eggs in a dorsal pouch. While most species undergo direct development, some deposit their tadpoles in water. Nocturnal and arboreal.

Gastrotheca cornuta — **Horned Marsupial Frog**

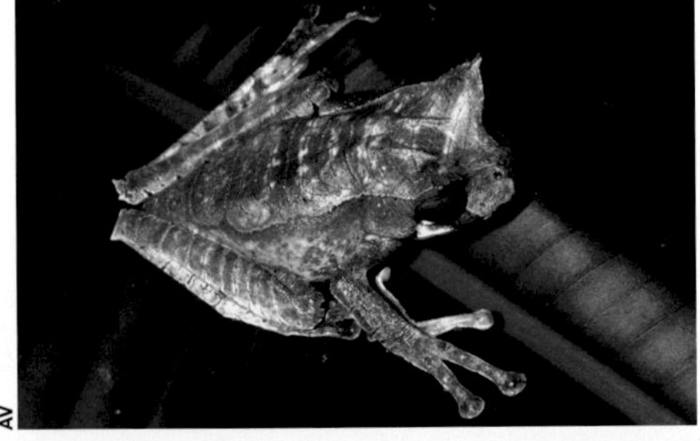

AV

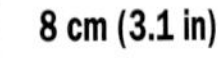

8 cm (3.1 in)

Note large triangular projections over the eyes. Females carry eggs and tadpoles within a dorsal pouch.

Hylidae (43 species)
Treefrogs

Treefrogs are nocturnal, arboreal frogs with well-formed finger and toe disks. Also note long legs. The majority of species have smooth skin while a few have rugose skin. Species in the genus *Agalychnis* have vertical pupils; all other species have horizontal pupils.

Agalychnis annae — **Golden-eyed Leaf Frog**

RDJ

8.4 cm (3.3 in)

Green overall with blue flanks. Golden eyes. Slender body. Formerly common in forest patches in the Central Valley but now restricted to less-developed areas.

Agalychnis callidryas — **Red-eyed Leaf Frog**

7.6 cm (3 in)

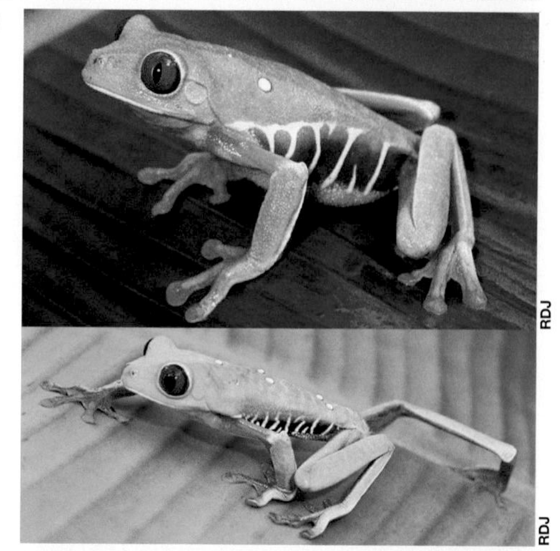

RDJ

RDJ

Yellow lines crisscross blue flanks; also note orange thighs, fingers, and toes; red eyes. A common nocturnal species.

Agalychnis (Phyllomedusa) lemur — **Lemur Treefrog**

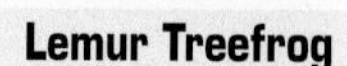

5 cm (2 in)

RDJ

AV

This fascinating frog is green by day, tan at night. Note how large eyes are oriented toward the side; also note vertical pupils. Lacks webbing on feet and hands. Slender body.

Agalychnis saltator — Jumping Leaf Frog

AV

6.3 cm (2.5 in)

Combination of red eyes and blue flanks and thighs distinguishes this species from other leaf frogs. Shows extensive webbing between fingers and toes. When breeding, jumping leaf frogs form groups of up to 200 individuals on vegetation above ponds.

Agalychnis spurrelli — Dark-eyed Leaf Frog

RDJ

8.9 cm (3.5 in)

Note yellow flanks, thighs, and venter. Has dark-red eyes. Extensive webbing between toes allows the frog to glide through the air. Nocturnal and arboreal.

Anotheca spinosa — Spiny-headed Treefrog

AV

7.3 cm (2.9 in)

Note striking brown and cream blotches on body and crownlike spines on head.

Cruziohyla (Agalychnis) calcarifer **Barred Leaf Frog**

7.6 cm (3 in)

RDJ

Dark-purple lines crisscross yellow flanks; also note yellow eyes. This species is a relatively uncommon inhabitant of the forest canopy.

Dendropsophus (Hyla) ebraccatus **Hourglass Treefrog**

3.3 cm (1.3 in)

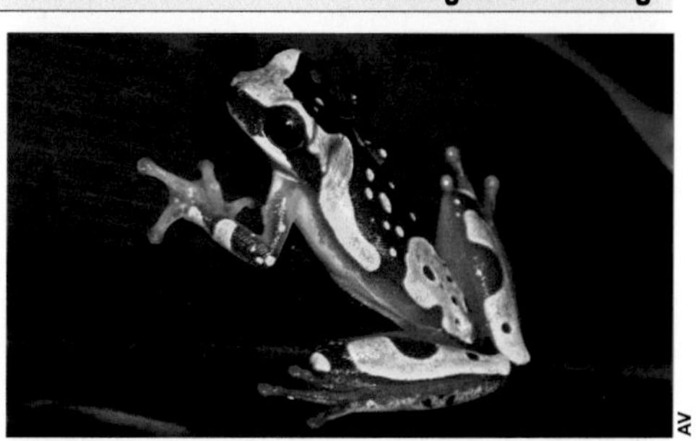

AV

Note yellow and brown curvy blotches (with yellow spots on brown areas and brown spots on yellow areas). Most common in lowland areas, generally near ponds.

Dendropsophus (Hyla) microcephalus **Yellow Treefrog**

3.1 cm (1.2 in)

DH

Yellowish overall; note darker flanks and venter. Very little webbing on fingers and toes. Makes a cricket-like call. A common frog, often found near ponds.

Dendropsophus (Hyla) phlebodes — **Veined Treefrog**

AV

2.7 cm (1.1 in)

Color varies from yellow to brown. Very little webbing on fingers and toes. A common frog in lowlands of the Caribbean slope, near ponds.

Duellmanohyla lythrodes — **Red-eyed Stream Frog**

Dark-green dorsum; note white line along each side of the body that separates dorsum from white venter. Also note red eyes and horizontal pupils. Similar to *D. rufioculis* (below), but distinguished from that frog by tympanum that is two-thirds the size of the eye. Also similar to *D. uranochroa* (p. 41), but distinguished from that species by dark-green dorsum and white venter. Reproduces on (and calls from) vegetation near streams.

3.3 cm (1.3 in)

Duellmanohyla rufioculis — **White-bellied Stream Frog**

RDJ

AV

3.8 cm (1.5 in)

Similar to *D. lythrodes* (above) in having dark-green dorsum and white venter, but distinguished by tympanum that is one-third the size of the eye. Note red eyes and horizontal pupil. *D. uranochroa* (p. 41) has light-green dorsum and yellow venter. Reproduces on (and calls from) vegetation near streams.

Duellmanohyla uranochroa **Costa Rica Stream Frog**

3.8 cm (1.5 in)

Note red eyes and horizontal pupils. Distinguished from *D. lythrodes* (p. 40) by light-green dorsum and yellow venter. Further distinguished from *D. rufioculis* (p. 40) by tympanum that is two-thirds the size of the eye. Reproduces on (and calls from) vegetation near streams.

Ecnomiohyla (Hyla) fimbrimembra **Highland Fringed-limbed Treefrog**

8.7 cm (3.4 in)

A brown frog with bronze eyes. Uses extensive webbing on large hands and feet to glide. Occurs at middle elevations, in forest canopy.

Ecnomiohyla (Hyla) miliaria **Lowland Fringed-limbed Treefrog**

10 cm (4 in)

Cream and brown blotches cover body. Note bronze eyes. Uses extensive webbing on large hands and feet to glide. Occurs in lowlands, in forest canopy.

Ecnomiohyla (Hyla) sukia **Shaman Fringed-limbed Treefrog**

8 cm (3 in)

Brown body with some light-brown blotches. This canopy inhabitant uses extensive webbing on large hands and feet to glide. Only member of its genus that has ossified shields on head and dorsum and in which the outer leg membrane reaches the toe.

Hyloscirtus (Hyla) colymba — Stream Treefrog

4.3 cm (1.7 in)

Dorsum green, sometimes with many tiny black spots; white venter; a yellow line runs from behind the eye, through the tympanum, and to the axilla. Note webbing between fingers and toes. Similar to *H. palmeri* (below). This rare frog occurs near fast-moving streams.

Hyloscirtus (Hyla) palmeri — Palmer's Treefrog

AV

5 cm (2 in)

Back uniformly green. Note yellow iris and horizontal pupil. Green bones visible through translucent-yellow venter. Has webbing on fingers and toes. Occurs near rocky streams.

Hypsiboas (Hyla) rosenbergi — Rosenberg's Treefrog

RDJ (RDJ)

9 cm (3.5 in)

Coloration varies from creamy yellow to creamy red. Most individuals show a black narrow line over the head and back. Webbing on large fingers and toes. Pointed nose.

Hypsiboas (Hyla) rufitelus **Red-webbed Treefrog**

5.3 cm (2.1 in)

juv.

RDJ (FMC)

RDJ

Note bright red webbing on toes and fingers; background color green (on some individuals brownish green). Has green bones. Juveniles up to 2.5 cm (1 in) in length lack red webbing and show a white dorsolateral line from nostril to groin.

Isthmohyla (Hyla) angustilineata **Pin-striped Treefrog**

3.8 cm (1.5 in)

Dorsum varies from dark green to brown; note narrow lateral stripe running from eye to hip. Venter is spotted. Lacks webbing between fingers. Uncommon. Lays eggs in forest ponds.

Isthmohyla (Hyla) calypsa **Lichen Treefrog**

3.8 cm (1.5 in)

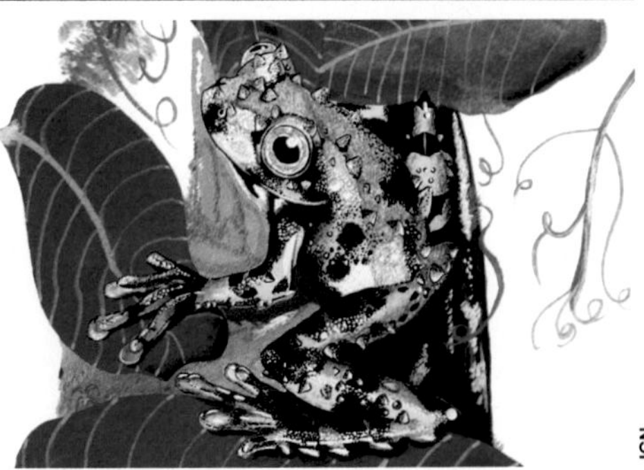

JCN

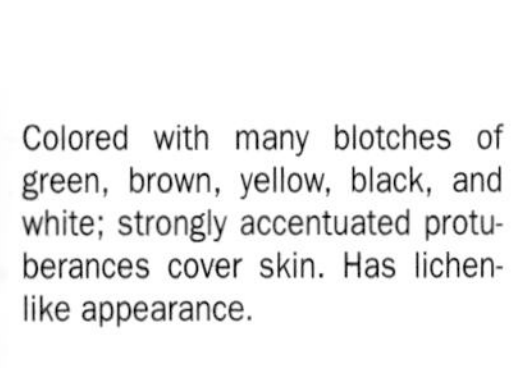

Colored with many blotches of green, brown, yellow, black, and white; strongly accentuated protuberances cover skin. Has lichen-like appearance.

Isthmohyla (Hyla) debilis — **Weak-voiced Treefrog**

Dorsum varies from pale green to brown. Venter white; note well-defined separation between dorsal and ventral colors. On some individuals, note a line from snout to tympanum (tympanum is very small). Iris orange. Uncommon. Occurs on vegetation high above streams.

3.1 cm (1.2 in)

Isthmohyla (Hyla) lancasteri — **Mottled Treefrog**

RDJ

4.1 cm (1.6 in)

Skin shows many blotches of green, brown, yellow, black, and white; skin surface resembles lichen. Similar to *I. calypsa* (p. 43).

Isthmohyla (Hyla) picadoi — **Picado's Treefrog**

Color overall varies from yellow to orange. Similar to *I. zeteki* (p. 46). Rare. Inhabits the forest canopy; most individuals have been found on bromeliads.

3.5 cm (1.4 in)

Isthmohyla (Hyla) pictipes — **Dark-footed Treefrog**

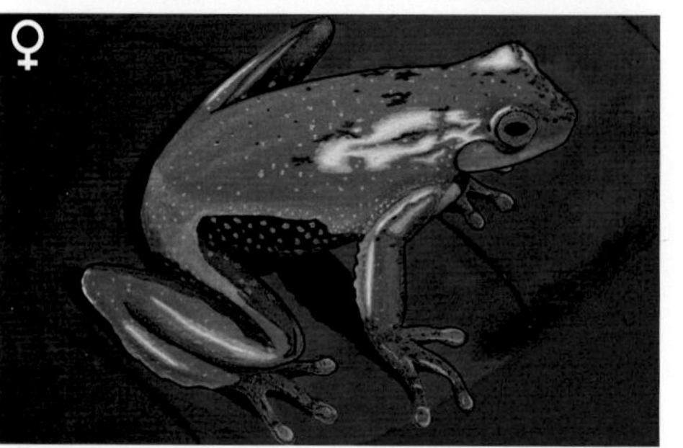

JCN

4.5 cm (1.8 in)

Dorsal color varies from individual to individual, from light green to brown to almost black, with small white or yellow spots; venter dark. Very small tympanum. Uncommon. Found close to the ground, near fast-moving streams.

Isthmohyla (Hyla) pseudopuma — Meadow Treefrog

5 cm (2 in)

RDJ

FMC

Variable in coloration, though generally brown with darker blotches all along the body; during breeding (in rainy season), males are bright yellow. Very common. Often found in ponds located in pastures or near roads.

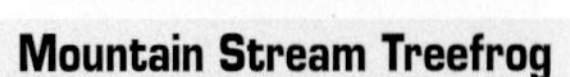

Isthmohyla (Hyla) rivularis — Mountain Stream Treefrog

3.6 cm (1.4 in)

FMC

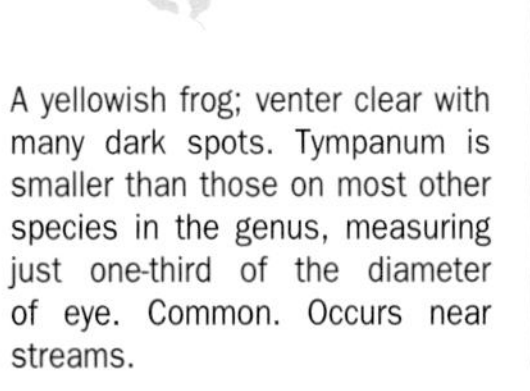

A yellowish frog; venter clear with many dark spots. Tympanum is smaller than those on most other species in the genus, measuring just one-third of the diameter of eye. Common. Occurs near streams.

Isthmohyla (Hyla) tica — Tico Treefrog

FMC

4.3 cm (1.7 in)

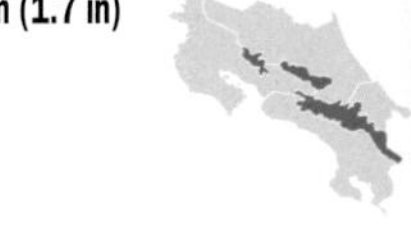

Greenish overall; blotchy tubercle-covered skin resembles lichen. Tympanum is fairly large (same diameter as that of eye).

Isthmohyla (Hyla) xanthosticta — Barva Treefrog

Has green dorsum and yellow venter. Known from only 2 specimens found in vicinity of Barva Volcano. Possibly extinct.

3 cm (1.2 in)

Isthmohyla (Hyla) zeteki — Zetek's Treefrog

RP

2.6 cm (1 in)

Varies in color from light green to yellow; eyes are large, brownish yellow. Has very slender body. Similar to *I. picadoi* (p. 44). Rare. Associated with bromeliads.

Osteopilus septentrionalis **Cuban Treefrog**

8.5 cm (3.3 in)

RG

INTRODUCED

Light brown overall, with dark-brown blotches. Note horizontal pupil, yellow iris. Little webbing on fingers and toes. Known only from Port of Limón area.

Ptychohyla (Hyla) legleri **Legler's Treefrog**

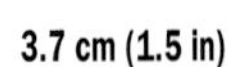

3.7 cm (1.5 in)

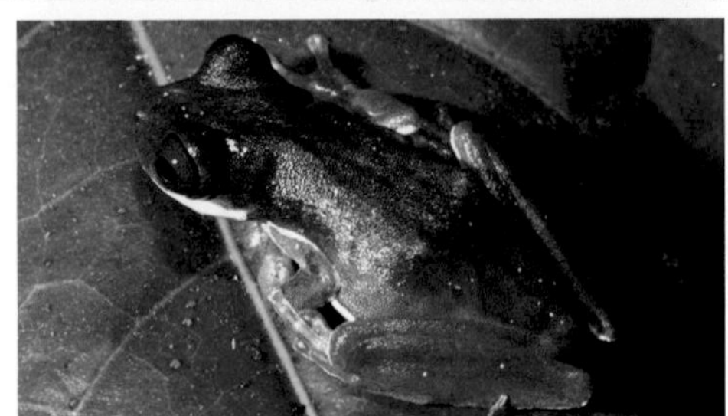

AV

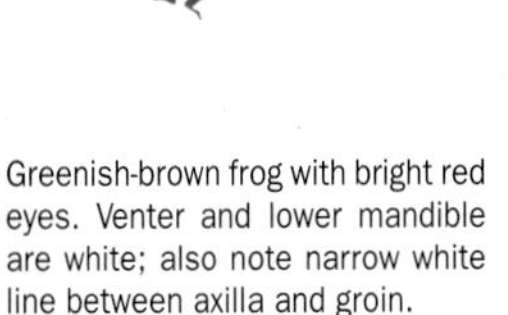

Greenish-brown frog with bright red eyes. Venter and lower mandible are white; also note narrow white line between axilla and groin.

Scinax boulengeri **Snouted Treefrog**

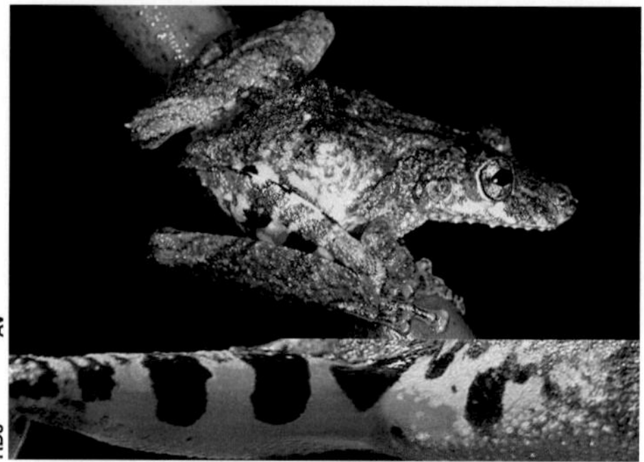

AV

RDJ

5.3 cm (2.1 in)

Note very pointed nose. Individuals are brown or green; also note black bars on posterior section of legs. Lacks webbing on hands.

Scinax elaeochrous **Narrow-headed Treefrog**

RDJ

FMC

4 cm (1.6 in)

Green, yellow, or brown frogs; some individuals have longitudinal blotches. Green bones. Lacks webbing on hands.

Scinax staufferi — **Stauffer's Treefrog**

3 cm (1.2 in)

AV

Yellowish-brown frog with longitudinal dark blotches. Has rough skin and a pointed nose. Lacks webbing on hands.

Smilisca baudinii — **Mexican Treefrog**

9 cm (3.5 in)

RDJ

RDJ

Has a roundish body. Coloration is highly variable, with some individuals a solid dark color and others covered in blotches. Often cream-brown color with dark blotches on dorsum and sides, and black bars that cross the lips. Also note transverse bars on hind legs. Forms loud aggregations near ponds at night.

Smilisca phaeota — Masked Treefrog

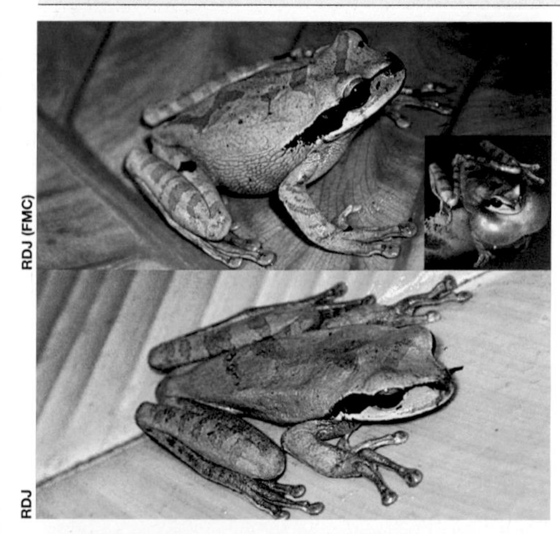

7.8 cm (3.1 in)

Individuals vary from creamy brown to green; note dark blotches running from the eye to the axilla. Forms loud aggregations near ponds at night.

Smilisca puma — Tawny Treefrog

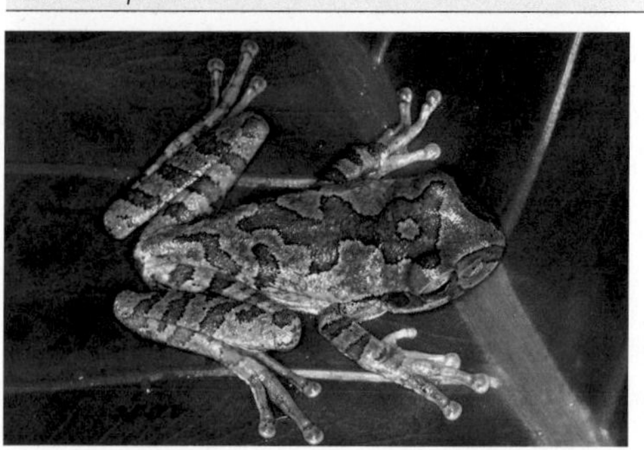

4.5 cm (1.8 in)

Varies greatly in coloration; some individuals are uniformly dark, others show dark blotches on a cream-yellow background. Skin generally smooth but note granular surface on venter. Has golden irises. Only slight webbing between fingers.

Smilisca sila **Pugnosed Treefrog**

6.2 cm (2.4 in)

Varies greatly in coloration; some individuals are uniformly dark, others show dark blotches on a cream-yellow background. Warty skin. Has brown irises. Only slight webbing between fingers.

Smilisca sordida **Drab Treefrog**

6.4 cm (2.5 in)

Varies greatly in coloration; some individuals are uniformly dark, others show dark blotches on a cream-yellow (to reddish-yellow) background. Smooth skin. Has brown irises. Note extensive webbing between fingers.

Tlalocohyla (Hyla) loquax **Swamp Frog**

AV
RDJ

4.2 cm (1.7 in)

Color varies from yellow to brown; shows dark-brown blotches on dorsum. Note bright red on toe webbing, thighs, groin, and axilla. Occurs near ponds.

Trachycephalus (Phrynohyas) venulosus **Pepper Treefrog**

RDJ

10 cm (4 in)

Note robust body and short legs Varies in color from yellow t brown; some individuals shov large longitudinal patches. Ski secretes irritating substance tha can induce sneezing in humans. O rainy nights, found on low veget tion near ponds.

Centrolenidae (13 species)
Glass Frogs

Glass frogs are characterized by translucent skin on venter. They have horizontal pupils. Glass frogs resemble treefrogs except that their eyes face forward to a greater degree than do those of treefrogs. Five genera occur in Costa Rica; in four genera, frogs have green bones, while members of *Hyalinobatrachium* have white bones. Males guard eggs, which are laid on underside of leaves that overhang water. These nocturnal and arboreal frogs are often found near streams.

Cochranella euknemos — **Cascade Glass Frog**

3.2 cm (1.3 in)

TL

Dorsum shows distinct pale spots. Note fleshy ridge along posterior margin of lower arm and lower leg.

Cochranella granulosa — **Granular Glass Frog**

3.4 cm (1.3 in)

RDJ

White dots on dorsum give this frog a granular appearance; many individuals also have black spots on dorsum. In profile, snout very pointed.

Espadarana (Centrolene) prosoblepon **Emerald Glass Frog**

RDJ

RDJ (FMC)

3 cm (1.2 in)

On many individuals, dorsum is speckled with small black spots. Nostrils located on high protuberances. Males have a humeral hook.

Hyalinobatrachium chirripoi **Chirripó Glass Frog**

Dorsum shows yellow spots. Extensive webbing on hands.

3 cm (1.2 in)

Hyalinobatrachium colymbiphyllum **Bare-hearted Glass Frog**

RDJ (RDJ)

2.5 cm (1 in)

Green dorsum shows yellow spots Nostrils elevated on high protube ance. Pericardium is translucent similar *H. talamancae* (p. 55) ha white pericardium.

Hyalinobatrachium fleischmanni — Fleischmann's Glass Frog

3 cm (1.2 in)

RDJ

Green dorsum shows many small yellow dots. Nostrils are slightly elevated.

Hyalinobatrachium talamancae — Talamanca Glass Frog

2.5 cm (1 in)

Similar to *H. colymbiphyllum* (p. 54) but with white pericardium (visible through the belly). Known only from Moravia de Chirripó.

Hyalinobatrachium valerioi — Reticulated Glass Frog

2.5 cm (1 in)

RDJ (RDJ)

On green dorsum, note yellow spots bordered by tiny black dots. Pericardium is white.

Hyalinobatrachium vireovittatum **Green-striped Glass Frog**

AV

2.5 cm (1 in)

Background color is green; on dorsum, note central green stripes bordered by yellow blotches.

Sachatamia (Cochranella) albomaculata **Spotted Glass Frog**

RDJ

2.5 cm (1 in)

Dorsum has many distinct yellow spots. Posterior side of each forearm has a fleshy fold.

Sachatamia (Centrolene) ilex **Ghost Glass Frog**

RDJ

3.2 cm (1.3 in)

Dorsum uniformly green; some individuals also show a few spots on the dorsum. On bizarre eyes, irregular black lines (on white background) surround pupil. Nostrils located on high protuberances. Note extensive webbing on hands.

Teratohyla (Cochranella) pulverata **Speckled Glass Frog**

3 cm (1.2 in)

RDJ (RDJ)

Dorsum with distinct light spots. Note fleshy ridge along posterior margin of lower arm and lower leg.

Teratohyla (Cochranella) spinosa **Spiny Glass Frog**

2.5 cm (1 in)

TL

Dorsal skin is lightly granular. Distinct spine occurs on the first digit (counted from the inside outward) of each hand.

Dendrobatidae (7 species)
Poison Frogs

The majority of these small frogs are extremely colorful. Their finger disks are divided by a longitudinal groove. Pupils are horizontal. Most species secrete toxic substances through their skin; rocket frogs (*Silverstoneia*) lack skin toxins. Poison frogs tend to be most active during the day and are generally found on the forest floor or on low-lying vegetation.

Dendrobates auratus — Green & Black Poison Frog

RDJ

RDJ

4.3 cm (1.7 in)

Combination of bright green and black patches is distinctive. Individuals from the Caribbean slope are mostly pastel green, while those from the Pacific are often metallic green and show more black. Very common in old cacao plantations.

Oophaga (Dendrobates) granulifera — Granular Poison Frog

RDJ

2.5 cm (1 in)

Red dorsum; green legs. Granular skin. Most common in shaded areas near small streams.

Oophaga (Dendrobates) pumilio — Strawberry Poison Frog

2.5 cm (1 in)

RDJ

RDJ

Individuals at higher elevations generally have a red dorsum and blue legs; those found near the coast tend to be reddish orange with black spots.

Phyllobates lugubris — Lovely Poison Frog

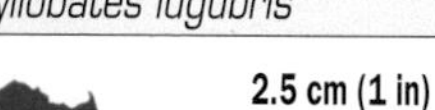

2.5 cm (1 in)

TL

On dark background, note bright yellow line (from nose to groin) on each side of the dorsum. Legs mottled with green, black, and gold. Furtive.

Phyllobates vittatus — Golfo Dulce Poison Frog

RDJ

3 cm (1.2 in)

On dark background, note bright orange line (from nose to groin) on each side of the dorsum. Legs mottled with green, black, and gold. Furtive.

Silverstoneia (Colostethus) flotator — Lowland Rocket Frog

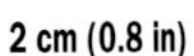

JPM

RDJ

2 cm (0.8 in)

Brown on most of body. On each side of body, note oblique white line from groin to eye; on individuals on Pacific slope, the line is complete, while on individuals on the Caribbean slope the line is interrupted. Color on appendages too variable to use as a way of distinguishing between the 2 populations. White venter. Makes short jumps, moves very quickly.

Silverstoneia (Colostethus) nubicola **Highland Rocket Frog**

2 cm (0.8 in)

Brown; on each side of body note 2 distinct white lines from groin to eye. Black venter. Makes short jumps, moves very quickly.

JCN

Aromobatidae (1 species)
Rocket Frogs

These small frogs are generally active during the day and are found on the forest floor. Their finger disks are divided by a longitudinal groove. Pupils are horizontal. Rocket frogs are very similar to the rocket frogs (genus *Silverstoneia*, p. 60 and above) of the poison frog family. This family is represented by a single species in Costa Rica.

Allobates (Colostethus) talamancae **Talamanca Rocket Frog**

2 cm (0.8 in)

FMC

Brown with a longitudinal white line from snout to axilla. White venter. Makes short jumps, moves very quickly.

Microhylidae (3 species)
Narrow-mouthed Toads

Narrow-mouthed toads are characterized by a small head and round robust body. Also note transverse fold behind the eyes, a feature that is unique to this family. Short legs. All Costa Rican species have small eyes with round pupils and are primarily nocturnal. Secretive and fossorial. Feeds on ants and termites.

Gastrophryne pictiventris — Southern Narrow-mouthed Toad

AV

4 cm (1.6 in)

Depending on individual, dorsum either dark brown or blue; also note bright white patches (or spots) on belly. Smooth skin.

Hypopachus variolosus — Sheep Frog

RDJ

5 cm (2 in)

Dorsum and legs are creamy brown; note dark-brown flanks.

Nelsonophryne aterrima — Black Narrow-mouthed Toad

7 cm (2.8 in)

TL

Uniformly blackish blue on dorsum and flanks. Very smooth skin.

Ranidae (5 species)
Pond Frogs

These aquatic or semiaquatic frogs have large eyes and tympanum, pointed heads, and long legs. Also note webbing between toes; distinguished from similar-looking frogs in the family Leptodactylidae (p. 33) by extensive toe webbing. Pupils are horizontal. Energetic jumpers. Active both during the day and at night.

Lithobates (Rana) forreri — Forrer's Leopard Frog

10 cm (4 in)

FMC

Varies from yellowish brown to green, with leopard-like brown blotches on dorsum, flanks, and legs. Pointed nose. One continuous skin fold on each side of dorsum, running from eye to groin. Mostly aquatic.

Lithobates (Rana) taylori — Taylor's Leopard Frog

RDJ

9 cm (3.5 in)

Varies from yellowish brown to green, with leopard-like brown blotches on dorsum, flanks, and legs. Pointed nose. One discontinuous skin fold on each side of the dorsum, running from eye to groin. Mostly aquatic.

Lithobates (Rana) vaillanti — Rainforest Frog

RDJ

12 cm (4.7 in)

Brown coloration on posterior region gradually changes to green toward front of body. Green to yellow lips. Pointed nose. Mostly aquatic.

Lithobates (Rana) vibicarius — Green-eyed Frog

RP

9 cm (3.5 in)

Green dorsum; flanks vary from dark brown to reddish brown. Yellow to white upper lip. Green irises. Black face mask.

6 cm (2.4 in)

Dark- to light-brown dorsum; dark flanks; and striking yellow, red, and black blotches on thighs and groin.

reptilia

Reptiles

227 species

Cheloniidae (4 species)
Hard Shelled Sea Turtles

This family includes all marine turtles except the Leatherback Sea Turtle (p. 69). Distinguishing features include a carapace and plastron covered with shields that are present throughout the life of the turtle, a carapace that is round (to oblong), and claws on the flippers. These sea turtles spend virtually their entire life in the ocean, coming ashore only to lay their eggs. Most nesting females return to the same beach where they were hatched, arriving at specific times of the year. After digging the nest, the female lays her eggs, covers them with sand, and then returns to the ocean. Nesting generally occurs at night.

Caretta caretta — Loggerhead Sea Turtle

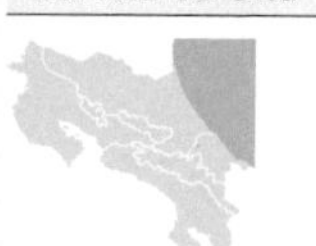

1 m (3.3 ft)

RDJ

On carapace, head, and limbs, background color is brown; overlaying reddish cast distinguishes it from other sea turtles. Only a few nests known from Tortuguero and Gandoca.

Chelonia mydas (agassizii) — Green Sea Turtle

LB

1 m (3.3 ft)

Gray-green dorsum; white sutures between shields in carapace, head, and limbs make green sea turtles easy to recognize. The first pair of lateral shields does not come into contact with the neck shield. This species is herbivorous after the third year, feeding mainly on marine algae. Tortuguero is one of the largest nesting beaches in the world for the species. The name *green turtle* comes from the edible green gelatinous substance obtained by scraping the internal surface of the plastron of slaughtered turtles.

Eretmochelys imbricata — Hawksbill Sea Turtle

RDJ

91 cm (3 ft)

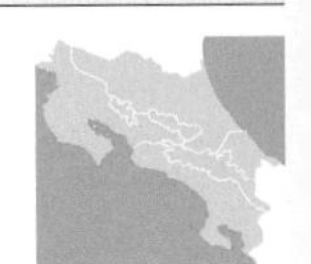

Dark-brown dorsum is beautifully marked with streaks of amber yellow and red; shields on the dorsum overlap; also note serrated edge of carapace. Esteemed with fervor equal to that of black coral, material from the shell (in Spanish, *carey*) is used (illegally) to craft artisanal and other products.

Lepidochelys olivacea **Olive Ridley Sea Turtle**

75 cm (2.5 ft)

FMC

Dorsum varies from gray-green to olive green. Carapace is as long as it is wide. Known for massive congregations (in Spanish, *arribadas*) of nesting females on Nancite and Ostional beaches.

Dermochelyidae (1 species)
Leatherback Sea Turtle

This family contains a single species of sea turtle, one of the largest of all living reptiles. The leatherback turtle has no claws on its flippers; the carapace is elongate. Unlike other sea turtles, the carapace and plastron lack shields and are covered instead with skin. Hatchlings are covered with small shields, which are lost after the first year. Diet is primarily jellyfish.

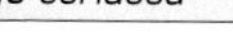

Dermochelys coriacea

Leatherback Sea Turtle

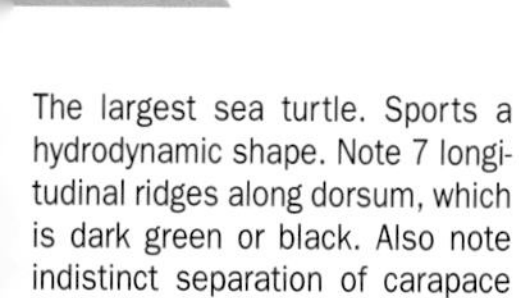

2 m (6.5 ft)

GS

The largest sea turtle. Sports a hydrodynamic shape. Note 7 longitudinal ridges along dorsum, which is dark green or black. Also note indistinct separation of carapace from rest of body.

Chelydridae (1 species)
Snapping Turtles

These aggressive aquatic turtles have a large head and a reduced plastron; also note large claws on the fingers and toes and a very long tail. Snapping turtles are opportunistic feeders of fish, frogs, reptiles, birds, and mammals, as well as aquatic plants. Largely nocturnal with some diurnal activity.

Chelydra acutirostris [serpentina] — South American Snapping Turtle

45 cm (18 in)

Long tail almost matches the length of its carapace; small carapace has distinct keels. Has a large head. On lower middle of chin, note 2 fleshy barbels (the exact function of which is still unclear, though they may aid in oxygen intake or serve as a sensory organ).

Kinosternidae (3 species)
Mud Turtles

Primarily aquatic. Generally found in slow-moving, often muddy, water. On most species, the hinged plastron allows the head, tail, and limbs to be enclosed within the shell for protection. Diet consists of both plant and animal matter. Expels a musky odor when disturbed.

Kinosternon angustipons — **Cross Mud Turtle**

13 cm (5 in)

RDJ

RDJ (RDJ)

Note smooth, dark-brown carapace. Head and neck are of same color (but of contrasting colors on 2 other *Kinosternon* species in Costa Rica). Reduced plastron resembles a cross. Lacks hooked beak.

Kinosternon leucostomum — **White-lipped Mud Turtle**

15 cm (6 in)

RDJ (RDJ)

Dark brown overall, but note yellow stripe on side of head and yellow lips; also note yellow venter. Its hinged plastron allows the turtle to seal off all soft body parts from potential predators. Has a hooked beak.

RDJ (RDJ)

17.5 cm (7 in)

Yellow to orange plastron. Red to yellow markings on face. Carapace with 3 longitudinal keels. Hinged plastron allows this turtle to seal its carapace. Has a hooked beak.

Emydidae (2 species)
Sliders

On sliders, the plastron is composed of 6 pairs of shields. Behind each eye, note yellow and orange stripes on a green background. Costa Rican members of this family are aquatic but often bask on logs or on the banks along lakes and rivers. Principally nocturnal, but are also seen during the day. Diet is mostly plant matter and, occasionally, invertebrates. Juvenile sliders are the most common turtles in the pet trade.

Trachemys emolli (scripta) **Nicaraguan Slider**

35 cm (14 in)

Bright yellow and green stripes run from eye to neck. Also note characteristic orange to red patch behind the eye. Legs green and yellow. Plastron yellow with green blotches along margin. Often seen sleeping on tree trunks over calm water.

FMC

Trachemys venusta (scripta) **Mesoamerican Slider**

35 cm (14 in)

Bright yellow and green stripes run from eye to neck. Also note characteristic yellow patch behind the eye. Legs green and yellow. Plastron yellow with green blotches along margin. Often seen sleeping on tree trunks over calm water.

RDJ

Geoemydidae (Bataguridae) (3 species)
Wood Turtles

On wood turtles, 6 pairs of shields form the plastron. Sides of neck usually show red spots on a yellowish background. Although most species are terrestrial (and found in forests), some species such as *Rhinoclemmys funerea* are primarily aquatic. Costa Rican species are largely diurnal and feed mostly on plant matter.

Rhinoclemmys annulata **Brown Forest Turtle**

20 cm (8 in)

JES

JES

Dark-brown dorsum sometimes shows fine yellow markings. Also note yellow longitudinal lines on neck. Legs dark. Plastron black with a prominent, yellow peripheral ring. Mostly terrestrial.

Rhinoclemmys funerea — **Black River Turtle**

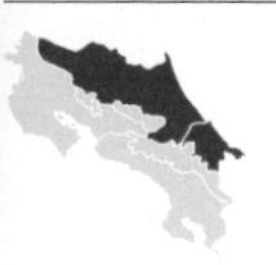

38 cm (15 in)

FMC

Black dorsum, neck, and face. Legs and plastron also dark. Aquatic.

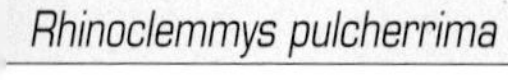

Rhinoclemmys pulcherrima — **Painted Wood Turtle**

20 cm (8 in)

RDJ (RDJ)

Distinctive dorsum shows reddish-brown and yellow markings. Neck is brown with yellow and red longitudinal lines. Light-brown legs. Yellow plastron has a peripheral ring, the colors of which are similar to those of a coral snake (p. 157).

SQUAMATA-SAURIA (73 species) Lizards

Corytophanidae (4 species)
Basilisk Lizards

Basilisks are lizards of medium size that have prominent head crests, long tails, long legs, and large eyes. Members of this family are diurnal. Species in the genus *Corytophanes* are arboreal; species in the genus *Basiliscus*, which are both terrestrial and arboreal, often perch on low branches. *Basiliscus* lizards are perhaps best known for their ability to run across the surface of water.

Basiliscus basiliscus — Common Basilisk

20 cm (8 in)

Extremely variable in coloration. Generally, light brown over most of body, with a wide yellow stripe from eye to tail and a second stripe (yellow or white) from nostril to shoulder, both intersected by black transverse bands; note that the stripes are usually bolder on males than on females, with some females showing little or no striping. Tail is twice as long as body. Males have an elevated crest on the head, trunk, and tail; on females and juveniles, crest (small) appears only on head—if at all. Runs on long hind legs. Found near water.

Basiliscus plumifrons — Emerald Basilisk

23 cm (9 in)

FMC

RDJ

Leaf-green color overall, with small blue and yellow spots along the body. Tail is twice as long as body. Males have an elevated crest on the head, trunk, and tail; on females and juveniles, crest (small) appears only on head—if at all. Runs on long hind legs. Found near water.

Basiliscus vittatus — Brown Basilisk

18 cm (7.1 in)

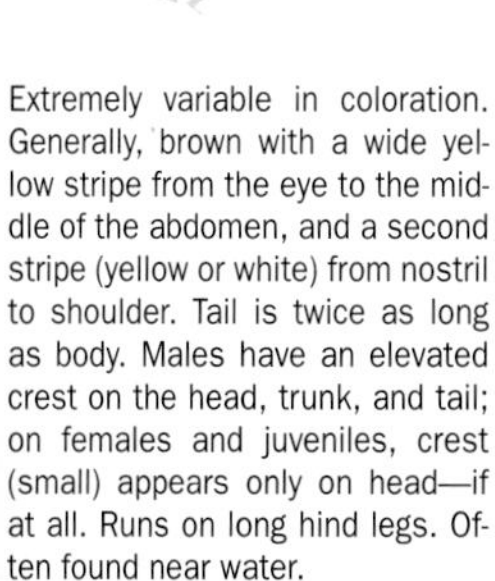

Extremely variable in coloration. Generally, brown with a wide yellow stripe from the eye to the middle of the abdomen, and a second stripe (yellow or white) from nostril to shoulder. Tail is twice as long as body. Males have an elevated crest on the head, trunk, and tail; on females and juveniles, crest (small) appears only on head—if at all. Runs on long hind legs. Often found near water.

RDJ

FMC

13 cm (5 in)

Able to change background color from brown to green. Two serrated crests on head join to form dome above neck. Long tail and legs. Arboreal, often found clinging vertically to trunks.

Iguanidae (3 species)
Iguanas

Iguanas are very large lizards with robust bodies and long tails. Also note long legs, long claws, and dorsal crests. Diet consists mainly of plant matter although juveniles often eat insects. Species in the genus *Ctenosaura* are terrestrial; those in the genus *Iguana* are mostly arboreal but can also be found on the ground. All are diurnal.

Ctenosaura quinquecarinata — Lesser Ctenosaur

18 cm (7.1 in)

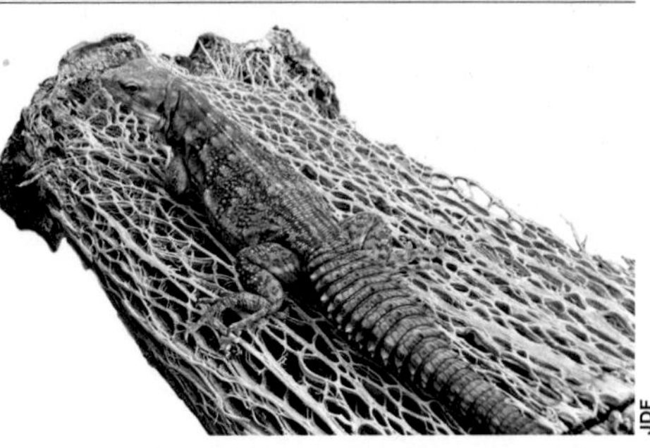

JDF

On adults, background color varies from gray to black; also note lateral black stripes. On adults and juveniles, tail is adorned with parallel rings of elevated spiny scales; between each of these parallel rings is a single fainter ring (two rings on *C. similis*, below) composed of small scales. On adults, a spiny dorsal crest runs from neck to base of tail; females have smaller dorsal crests than males. Juveniles are leaf green with black dots and a gray tail.

Ctenosaura similis — Spinytail Iguana

49 cm (19 in)

RDJ RDJ RDJ

On adults, background color varies from gray to black; also note lateral black stripes. On adults and juveniles, tail is adorned with parallel rings of elevated spiny scales; between each of these parallel rings are two fainter rings composed of small scales. On adult male, note spiny dorsal crest that runs from the neck all the way to the tail (female has reduced crest). Juveniles are leaf green with black dots and a gray tail.

60 cm (2 ft)

Males, females, and juveniles each distinct in color and morphology. Adult males vary from grayish green to orange and have a very large dewlap. Adult females are grayish green and lack dewlap. Juveniles are leaf green. Middorsal crest is made of long spiny individual scales; on juveniles, middorsal crest is very small and lacks the long spines. Tail, twice as long as the body, has many wide black rings. Also note prominent smooth round scale below the ear.

Phrynosomatidae (3 species)
Spiny Lizards

Spiny lizards have spiny robust bodies that range in size from small to medium. Also note large heads (without crests), legs of moderate length, and long fingers. These diurnal lizards are both terrestrial and arboreal.

Sceloporus malachiticus — **Green Spiny Lizard**

10 cm (4 in)

♂ ♀

FMC

AV

Males are malachite green; females are mottled with green, brown, and cream. Note rough skin. From neck to thighs, middorsal scale rows count fewer than 40 (see *Sceloporus variabilis*, p. 82). Mostly found on trees, rocks, and walls of yards and gardens.

Sceloporus squamosus — **Bocourt's Swift**

5.8 cm (2.3 in)

TL

Dorsum is brown. Head coloration varies from brown to reddish brown. From neck to thighs, middorsal scale rows count fewer than 40 (see *Sceloporus variabilis*, p. 82). Mostly terrestrial.

Sceloporus variabilis **Rose-bellied Swift**

FMC (RDJ)

7.6 cm (3 in)

Dorsum is brown. A light stripe runs from each eye to base of tail. Venter is salmon pink. On reproductive males, note blue ventral patches. From neck to thighs, middorsal scale rows count more than 40 (see *S. malachiticus* and *S. squamosus*, p. 81). Terrestrial.

Dactyloidae (26 species)
Anoles

Anoles are small, slender lizards with long legs and tails and expanded finger pads. Adult males have expandable dewlap that bears species-specific coloration. Many anole species have the ability to change color and pattern. This abundant lizard group is the one most commonly observed in Costa Rica. Recent work has proposed splitting certain Costa Rican anoles into separate species. Mostly arboreal.

Anolis (Norops) altae **Montane Anole**

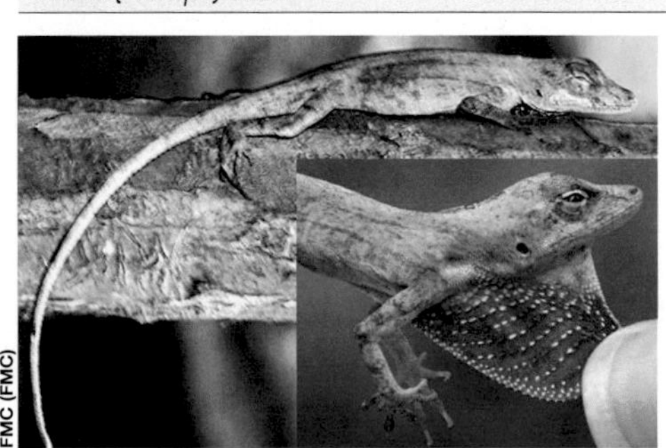

FMC (FMC)

5 cm (2 in)

Dewlap orange. Color overall varies from brown to light brown. Short legs. Similar to *A. kemptoni* (p. 87).

Anolis (Norops) aquaticus **Water Anole**

7.6 cm (3 in)

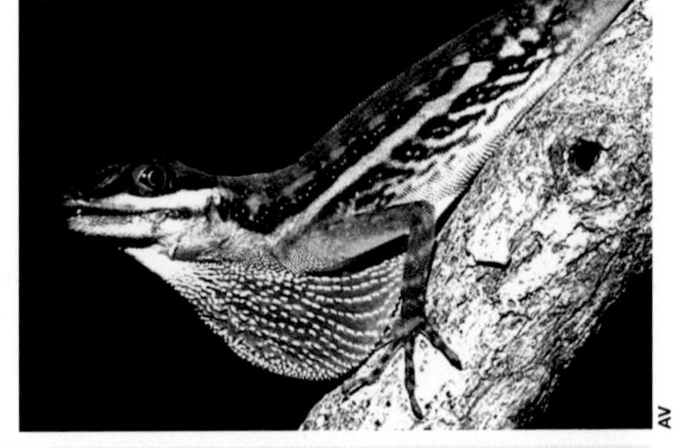

AV

Large dewlap is orange with yellow stripes. On dorsum, background color (brown to green) is intersected by light bands. A white line runs below each eye (which is blue); also note white line between shoulder and waist. White venter. Found on rocks next to streams. Able to walk underwater.

Anolis (Norops) biporcatus **Green Tree Anole**

10 cm (4 in)

RDJ

RDJ (AV)

Dewlap with blue center, red margin, and white base. Dorsum is leaf green, but this anole can change its color to brown in a few seconds.

Anolis (Norops) capito **Pug-nosed Anole**

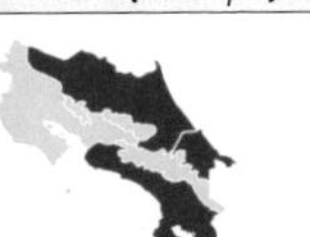

10 cm (4 in)

RDJ

RDJ

Dewlap greenish yellow and small. On female and male, note long legs and tail. On male, coloration resembles moss on tree trunk; head is short (but note that, in profile, head shows "crown"). Female is brown, with dorsal stripes.

Anolis (Norops) carpenteri — Little Green Anole

RDJ

4.3 cm (1.7 in)

Dewlap bright orange. Dorsum is leaf green but transforms itself to brown in a few seconds. Female has orange spot on throat.

Anolis (Ctenonotus) cristatellus — Crested Anole

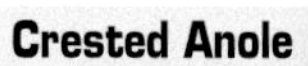

AV

7 cm (2.8 in)

INTRODUCED

Dewlap yellowish green with orange margins; only male anole with caudal crest. Color varies from brown to cream.

Anolis (Norops) cupreus — Dry Forest Anole

RDJ

5.6 cm (2.2 in)

Dewlap reddish orange, with anterior third white. Dorsum varies from brown to light brown. Commonly occurs in dry habitats, on low perches near the ground.

Anolis (Dactyloa) frenatus — **Spotted Tree Lizard**

14 cm (5.5 in)

Dewlap creamy white. Green background mottled with round, dark-brown blotches. Very long tail. Arboreal.

RDJ

Anolis (Norops) fungosus — **Mottled Anole**

5 cm (2 in)

Dewlap red with lines composed of white spots. Coloration resembles fungus on wood. Short legs. Known from only 5 individuals collected from premontane rainforest.

Anolis (Norops) humilis — **Ground Anole**

5 cm (2 in)

RDJ (RDJ)

Dewlap bright red with yellow margin. Dorsum varies from brown to dark brown. Only anole with a pocket hole in its axilla (the function of this pocket is as yet undetermined). Very common in wet habitats of lowlands, usually on forest floor.

Anolis (Dactyloa) ibanezi (chocorum) — **Chirping Anole**

7.6 cm (3 in)

Dewlap orange with green lines. Male and female green with dark band on the sides; both have long legs. Arboreal. Uncommon, though occasionally seen near Bribrí.

Anolis (Dactyloa) insignis — Giant Banded Anole

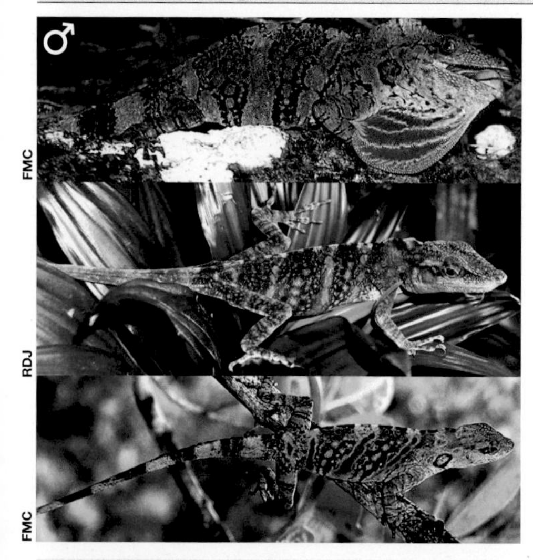

16.5 cm (6.5 in)

Dewlap dark red. A large anole. Shows a pattern of green and gray bands; capable of changing color of bands (green to brown, gray to orange) very quickly (in less than 60 seconds). Very long tail. Short legs. Note round patch behind ear.

Anolis (Norops) intermedius — Gray Lichen Anole

5 cm (2 in)

Dewlap white. Dorsal pattern variable; individuals show a mixture of colors, including dark brown, olive brown, light brown, and white. Note large yellow scales on lateral surfaces. Throat and most of ventral area white. Has short legs. Some females may have broad, tan middorsal stripe.

Anolis (Norops) kemptoni (pandoensis) — **Cerro Pando Anole**

5 cm (2 in)

Dewlap pink with orange margin. Dorsum color varies from brown to light brown. Short legs. Similar to *A. altae* (p. 82).

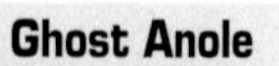

Anolis (Norops) lemurinus — **Ghost Anole**

7 cm (2.8 in)

Dewlap dark red, small. This anole varies dramatically from one individual to the next; some are a solid color, others a mix of colors; and, given that individuals are capable of quickly changing appearance, it is impossible to describe a "representative" species. Note, however, large yellow scales on lateral surfaces. Some females have broad, tan middorsal stripe or diamond-shaped marks on dorsum.

RDJ (AV)

JVH

Anolis (Norops) limifrons — **Slender Brown Anole**

5 cm (2 in)

Dewlap white with orange spot at base. Very slender anole. Has light-brown dorsum, white ventral areas, and banded tail. A common anole in Costa Rica.

FMC

Anolis (Norops) lionotus (oxylophus) **Stream Anole**

AV

8.9 cm (3.5 in)

Dewlap dark orange. Dorsum a uniform dark brown. Note copper irises. Conspicuous white lateral line runs from upper lip, through area above shoulder, to groin. Generally found next to lowland streams.

Anolis (Dactyloa) microtus **Small-eared Anole**

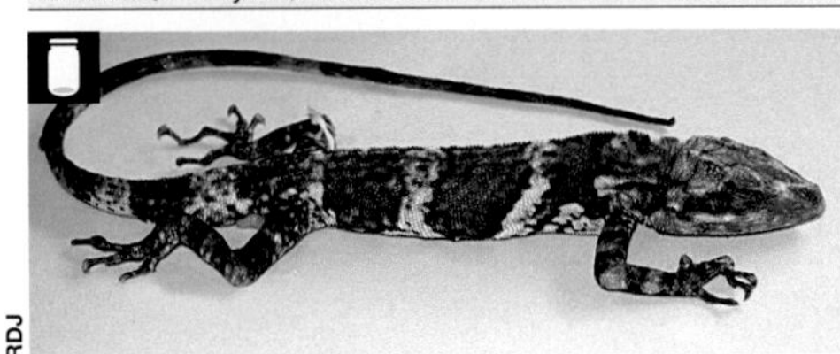

RDJ

11 cm (4.3 in)

Dewlap pink with white lines. Banded pattern can change coloration in less than a minute. Colors of bands vary and can be green, various shades of brown, or cream. Prominent white stripe behind ear. Long tail. Short legs.

Anolis (Norops) pachypus **Little Brown Anole**

Dewlap reddish orange. Dorsum is uniformly dark brown. Pattern on head of short white lines radiating from eye is somewhat similar to that of *A. tropidolepis* (p. 89).

5 cm (2 in)

Anolis (Norops) pentaprion **Lichen Anole**

RDJ (RDJ, FMC)

7.6 cm (3 in)

Dewlap varies from red to purple. Dorsum color pattern resembles lichen on wood. Short legs. Arboreal; tends to position itself flat against tree bark.

Anolis (Norops) polylepis — Jumping Anole

5 cm (2 in)

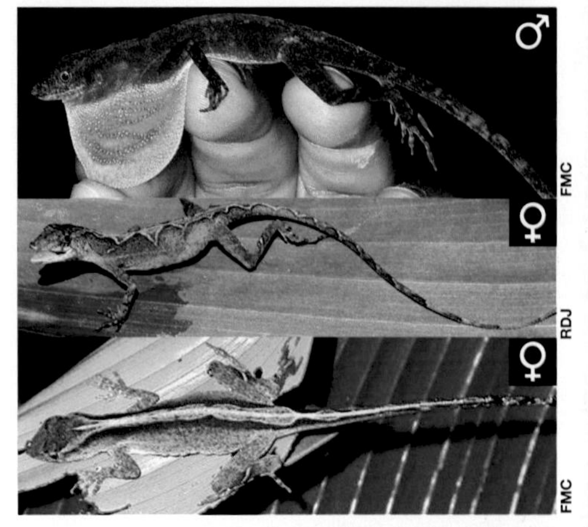

Dewlap orange. Dorsum generally uniformly light brown, though some females also show broad, tan middorsal stripe or diamonds. Slender with long legs. Jumps from branch to branch—or between branch and ground—much more frequently than other anoles. Most commonly found on forest floor and low perches.

Anolis (Norops) townsendi — Coco Island Anole

5 cm (2 in)

Dewlap amber. Olive-brown dorsum. Short legs. Endemic to Cocos Island—and it is the only anole that occurs there.

Anolis (Norops) tropidolepis — Cloud Forest Anole

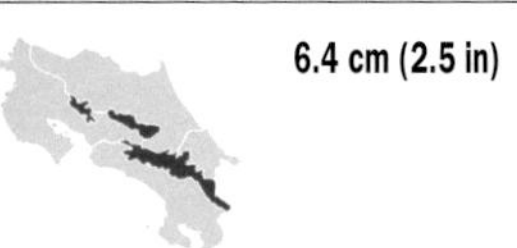

6.4 cm (2.5 in)

Dewlap red, small. Note yellow irises. Dark-brown dorsal pattern with some lighter patches; some females also show broad, tan middorsal stripe or diamonds. Distinctive white stripe runs from beneath the eye to below the shoulder (see *A. pachypus*, p. 88). Found on forest floor or on low perches.

Anolis (Norops) unilobatus (sericeus) **Moist Forest Anole**

FMC

5 cm (2 in)

Striking dewlap is yellowish orange with purple spot. A slender anole. On dorsum, creamy background with fine reticulations resembles bark. Some females have broad, tan middorsal stripe. Arboreal.

Anolis (Norops) vociferans **Vocal Anole**

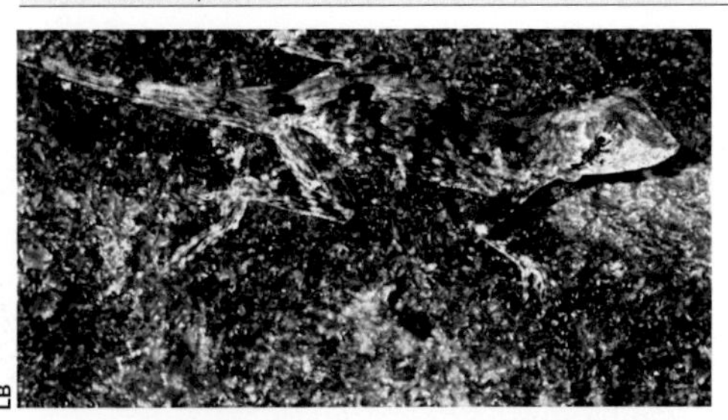

LB

5.8 cm (2.3 in)

Dewlap red. Dorsal pattern varies from gray to brown; resembles lichen. White venter. Short legs. Known from only a few individuals found at 2 locations in the western Talamanca region.

Anolis (Norops) woodi **Blue-eyed Anole**

FMC (FMC)

10 cm (4 in)

Color of dewlap can change from brown to orange in just a few seconds. Note bright blue eyes. Dorsal pattern of irregular blotches of brown, green, cream, gray, and black; resembles moss on wood. Long legs; long tail. Arboreal.

Polychrotidae (1 species)
Canopy Lizards

Canopy lizards are diurnal and occur in the upper canopy. Note long legs, very long tail, and angular head. The eyes are located on cones that can be rotated independently, in a fashion similar to that of true chameleons. Like the anoles, canopy lizards have dewlaps and can change colors.

Polychrus gutturosus **Canopy Lizard**

17.5 cm (7 in)

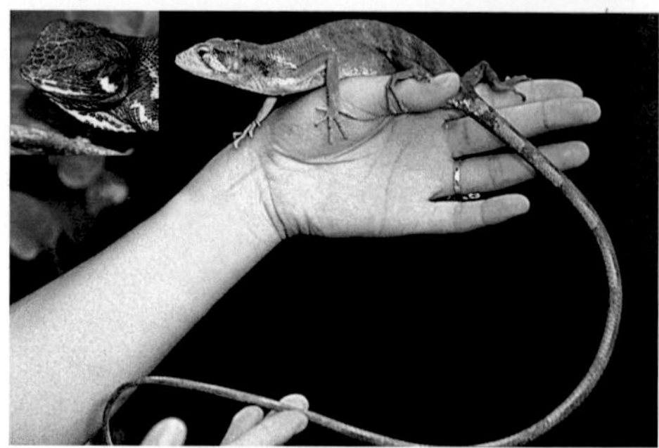

RDJ (RDJ)

Small dewlap is magenta, with lines formed by large green scales. Dorsum coloration varies from bright green to brown. Extremely long tail. No crests. Large scale under the ear.

Geckkonidae (12 species)
Geckos

Geckos are small- to medium-size lizards. Several species show noticeable digital pads on both fingers and toes that enable them to stick to smooth surfaces such as walls and ceilings. They have relatively thick tails. With the exception of species in the genus *Coleonyx*, all geckos in Costa Rica lack movable eyelids. Most species are primarily nocturnal. Some species vocalize.

Coleonyx mitratus — Tropical Banded Gecko

RDJ

10 cm (4 in)

On dorsum, wide reddish-brown bands alternate with narrow creamish-yellow bands. Note vertical pupils and movable eyelids. Walks with trunk elevated above the ground—only gecko in Costa Rica to do so. Digital pads not expanded.

Gonatodes albogularis — Yellow-headed Gecko

RDJ (RDJ)

RDJ

5 cm (2 in)

On males, note bright yellow (or orange) head and black body with lateral blue marks. Females and juveniles have irregular brown splotches that suggest the surface of wood. All individuals have white tail tips. Digital pads not expanded. Arboreal and diurnal.

Hemidactylus frenatus — Common House Gecko

6.5 cm (2.6 in)

RDJ (RDJ)

INTRODUCED

Tan to gray coloration. On chin, second submental scales touch infralabial scales (see photo insert). Has expanded digital pads. Makes chirping calls. Mostly active at night. Very common around human habitations, where often seen catching insects near lights.

Hemidactylus garnoti — Indopacific Gecko

6.5 cm (2.6 in)

RDJ (RDJ)

INTRODUCED

Tan to gray coloration. On chin, tiny scales separate second submental scales and infralabial scales (see photo insert). Flattened tail is serrated on the sides. Has expanded digital pads. Makes chirping calls. Generally active at night. Occurs in or near human habitations, where often seen catching insects near lights. Parthenogenetic. Less common than *H. frenatus* (above).

Lepidoblepharis xanthostigma — Yellow-spotted Gecko

4.5 cm (1.8 in)

AV

Dark-brown body with tiny yellow spots; also note yellow stripe behind the eye and long reddish tail. Digital pads not expanded. Diurnal. Moves quickly through leaf litter.

Lepidodactylus lugubris **Asian Gecko**

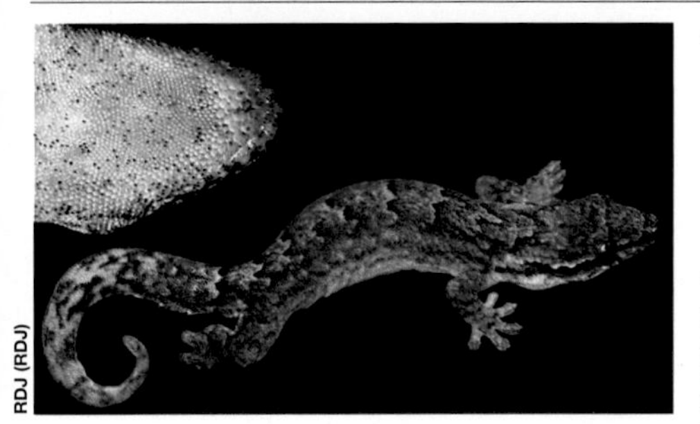
RDJ (RDJ)

4.5 cm (1.8 in)

INTRODUCED

Yellowish tan to gray coloration. Tail much thicker than tail on *H. frenatus* (p. 93) and *H. garnoti* (p. 93); also lacks the large submental scales found on both species (see photo insert). Has expanded digital pads. Parthenogenetic. Makes chirping calls. Found in human habitations. Mostly active at night, catching insects near lights. In Costa Rica, this species is less common than either *H. frenatus* or *H. garnoti*.

Phyllodactylus tuberculosus **Leaf-toed Gecko**

RDJ

8 cm (3.1 in)

Dorsum shows distinctive pattern of multiple dark elongated spots on a creamy background. Also note prominent tubercles on back, thick tail, and wide digital pads.

Sphaerodactylus graptolaemus **Western Pygmy Gecko**

RDJ

3.5 cm (1.4 in)

Note dark longitudinal stripes on a creamy background. Has no eyelids. Head covered with small granular scales. Digital pads not expanded. Diurnal forest floor inhabitant.

Sphaerodactylus homolepis **Eastern Pygmy Gecko**

3.5 cm (1.4 in)

Yellowish head shows dark stripes or spots. Head covered with small granular scales. Digital pads not expanded. Adults are striped or spotted. Juveniles with dark crossbands on a creamy background. Diurnal. Generally found on forest floor but sometimes ventures into trees.

Sphaerodactylus millepunctatus **Spotted Pygmy Gecko**

3.5 cm (1.4 in)

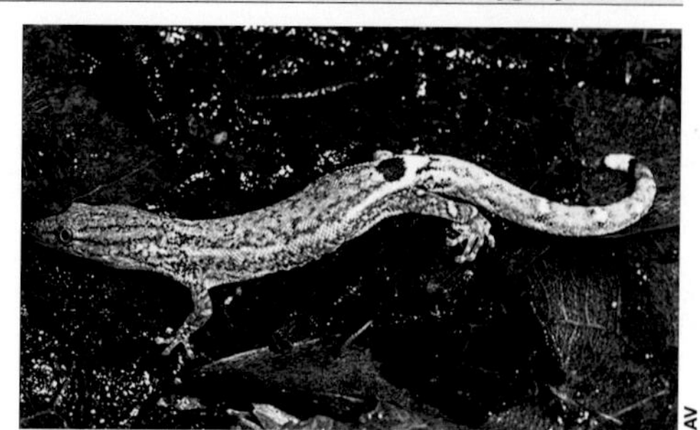

Dorsum yellowish with many small, dark spots. Head covered with small granular scales. Digital pads not expanded. Diurnal forest floor inhabitant.

Sphaerodactylus pacificus **Coco Island Pygmy Gecko**

5 cm (2 in)

Digital pads not expanded. Head covered with small granular scales. Diurnal forest floor inhabitant. The only endemic gecko on Coco Island.

Thecadactylus rapicauda **Turnip-tail Gecko**

RDJ

12 cm (4.7 in)

Note dark streaks on a light background, suggestive of the surface of bark. Also note large eyes and vertical pupils. Tail base is very thick. Has large, expanded digital pads. The largest gecko that occurs in Costa Rica. Arboreal and nocturnal.

Xantusiidae (2 species)
Night Lizards

Night lizards have boxy bodies, short legs, and a short tale. They lack eyelids. Also note that they have granular scales along the body and large scales on the head. Terrestrial and semifossorial, they are often found in leaf litter or under decaying logs. Costa Rican species are parthenogenetic, although males of *Lepidophyma flavimaculatum* do occur in northern Costa Rica. Nocturnal.

Lepidophyma flavimaculatum **Tropical Night Lizard**

RDJ

10 cm (4 in)

Dorsum dark black with multiple yellow spots; throat is pale. Head covered with smooth scales; body covered with granular scales.

Lepidophyma reticulatum — Reticulated Night Lizard

10 cm (4 in)

TL

Dorsum dark black with multiple yellow spots; throat shows pale roundish blotches. Head covered with smooth scales; body covered with granular scales.

Scincidae (3 species)
Skinks

Skinks have slender bodies with round shiny scales and short legs. Most species are easily recognized by their glossy scales and streamlined body. They move very quickly. Costa Rican skinks are diurnal; they occur both on the forest floor and beneath it.

Mabuya unimarginata — Bronze-backed Skink

10 cm (4 in)

RDJ

Dorsum metallic bronze; dark-brown flanks bordered below by white line. Diurnal. Terrestrial. Very fast-moving. See *Sphenomorphus cherriei* (p. 98).

Mesoscincus [Eumeces] managuae — Forest Skink

12 cm (4.7 in)

TL

On dorsum, note many black longitudinal lines on a light-brown background. Head has large scales; rest of the body has smooth cycloid scales. Uncommon.

Sphenomorphus cherriei — Striped Litter Skink

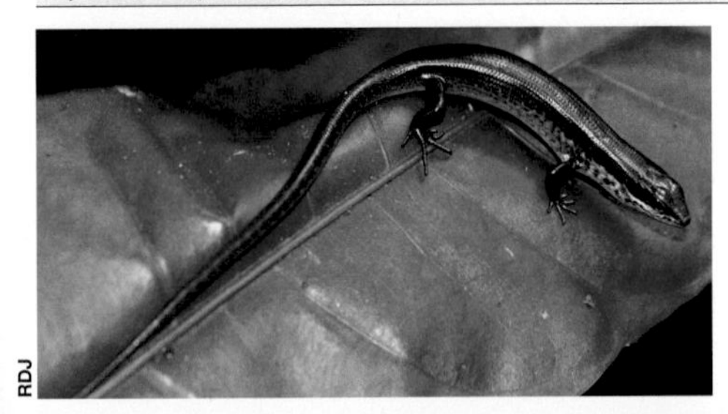
RDJ

6.4 cm (2.5 in)

Similar to *Mabuya unimarginata* (p. 97) but white line below flanks is wider; also distinguished by multiple black dots (mostly from shoulder to mouth). Moves very fast through leaf litter.

Teiidae (6 species)
Teiids

Teiids are robust lizards with pointed heads and large eyes. Their heads are covered with large plates; their bodies bear much smaller scales. Costa Rican species have long bodies and long, evenly tapered, whiplike tails. They are fast but, when not in motion, habitually bask in sunny spots. Terrestrial and diurnal.

Ameiva ameiva — Tropical Ameiva

FMC

19 cm (7.5 in)

Robust, large body; large triangular head; and long tail. Coloration varies with sex and age: adult males show green or yellow spots on a light-brown background; on females and juveniles, dorsum has light-brown background with 4 green or yellow stripes.

Ameiva festiva — Central American Ameiva

12.7 cm (5 in)

Robust, large body; large triangular head; and long tail. On all individuals, note 6 rows of elongated yellow dots running from head to tail. Middorsal area dark in adult males, white in adult females, blue in juveniles. In juveniles, tail also blue.

Ameiva leptophrys — Reticulated Ameiva

13 cm (5.1 in)

Robust, large body; large triangular head; long tail. On bronze-brown dorsum, note dark lateral stripe bordered above by yellow dots. Middorsal area crisscrossed by multiple black lines.

Ameiva quadrilineata — Four-lined Ameiva

9 cm (3.5 in)

Robust body; large triangular head; and long tail. Dorsum has brown background interrupted by 4 yellow longitudinal lines. Middorsal area light brown.

Ameiva undulata — **Barred Ameiva**

RDJ

13 cm (5.1 in)

Robust, large body; large triangular head; and long tail. Dorsum has brown background interrupted by a light-blue upper lateral stripe; flanks crisscrossed by black lines.

Cnemidophorus (Aspidoscelis) deppii — **Lined Racerunner**

AV

9 cm (3.5 in)

Robust body; large triangular head; and long tail. Dorsum has reddish-brown background interrupted by 10 yellow longitudinal lines. Bluish venter.

Gymnophthalmidae (6 species)
Microteiids

Microteiids have pointed heads that are covered with large scales. On most species in this family, body scales are square (to rectangular). They are mostly diurnal; they spend most of their time on the forest floor but are also fossorial. Many species have elongated bodies with short to very short legs. Microteiids scurry quickly through leaf litter in search of food.

Anadia ocellata — **Bromeliad Lizard**

7.5 cm (3 in)

AV

Coloration above varies from brown to golden brown (some individuals show black spots); white venter; on sides, note white spots (those near head encircled by black). Has a pointed head. Note squarish scales on body. Long tail measures two-thirds of the total length. Arboreal.

Bachia blairi (pallidiceps) — **Earless Lizard**

6 cm (2.4 in)

RDJ

Varies in color from dark gray to black. Long tail measures two-thirds of the total length. Extremely small legs give it a snakelike appearance. Squarish scales form rings around the body. Fossorial.

Gymnophthalmus speciosus — Spectacled Lizard

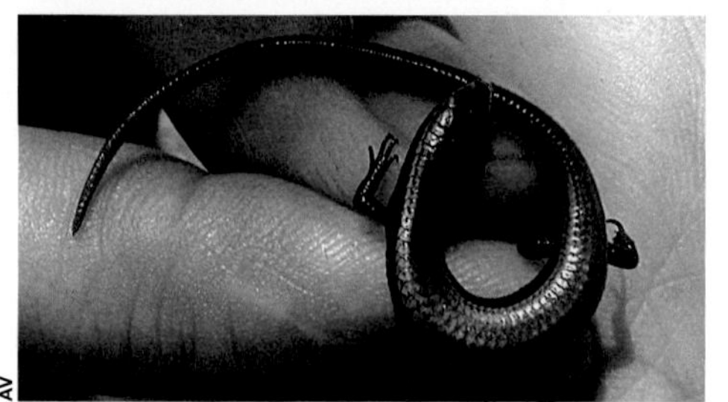

AV

4.5 cm (1.8 in)

On bronze dorsum note roundish scales. Tail is red and very long, measuring two-thirds of the total length. Short legs. Diurnal. Moves quickly through leaf litter.

Leposoma southi — Keeled Litter Lizard

RDJ (FMC)

4 cm (1.6 in)

Brown dorsum bordered on each side by cream line that runs from neck to tail. Scales on dorsum and head are squarish and keeled. Very long tail measures two-thirds of the total length. Found primarily in leaf litter.

Potamites (Neusticurus) apodemus — Water Lizard

TL

6 cm (2.4 in)

Brown dorsum. Scales on dorsum are squarish and keeled. Very long tail measures two-thirds of the total length. Semiaquatic.

Ptychoglossus plicatus **Largescale Lizard**

6.5 cm (2.6 in)

FMC

On brown dorsum, scales have noticeable keels; on head, scales are smooth; pale ventral scales are rectangular. Diurnal. Moves quickly through leaf litter.

ANGUIDAE (7 species)
Alligator Lizards

s their name suggests, alligator lizards have large mouths and bodies and short legs. Many pecies are colorful. Costa Rican members of this family are diurnal and have glossy scales nd moderate to very long tails. Though generally terrestrial, they are also often found hiding in af litter or under bark.

Celestus cyanochlorus **Green-bellied Caiman Lizard**

10 cm (4 in)

FMC

Dorsum varies from brown to bronze, uniformly covered with small black dots; unlike *C. orobius* (p. 104), has no lateral bars. Also note green venter and reddish tail. Has rounded, shiny, smooth scales. Neck has same width as head and body. Claws completely exposed.

Celestus hylaius **Spotted Caiman Lizard**

TL

10 cm (4 in)

Dorsum varies from brown to bronze, uniformly covered with small black dots that seem to form lines. Also note green venter and reddish tail. Has rounded, shiny, smooth scales. Neck has same width as head and body. Claws completely exposed.

Celestus orobius **Barred Caiman Lizard**

8 cm (3.1 in)

Dorsum varies from brown to bronze, uniformly dotted with small black dots; has pale vertical bars on sides. Also note green venter and reddish tail. Has rounded, shiny, smooth scales. Neck has same width as head and body. Claws completely exposed.

Coloptychon rhombifer **Isthmian Alligator Lizard**

AV

22 cm (8.7 in)

Brown dorsum with dark crossbands that form rhomboida figures. Head clearly differentiate from neck and trunk. Long tail Squarish scales. Claws complete exposed. Known from only specimens.

Diploglossus bilobatus **Green Galliwasp**

10 cm (4 in)

Dorsum varies from brown to bronze; green on sides with black blotches. Has rounded, shiny, smooth scales. Neck has same width as head and body. Claws exposed only at the very tips.

Diploglossus monotropis **Red Galliwasp**

20 cm (7.9 in)

Dorsum dark brown to black; red on sides with black blotches and vertical lines. Has rounded, shiny, smooth scales. Neck has same width as head and body. Claws exposed only at the very tips.

Mesaspis monticola **Tropical Alligator Lizard**

9 cm (3.5 in)

Strongly keeled dorsum. Long robust tail. On males, dorsum green; on females and juveniles, brown; on all individuals, dorsum covered with multiple black spots. Occurs at high elevations. Often found under rocks.

SQUAMATA-SERPENTES (138 species) Snakes

Anomalepididae (3 species)
Wormsnakes

Members of the family Anomalepididae (as well as those in the families Leptotyphlopidae and Typhlopidae, both p. 107) are slender burrowing snakes that are placed in the infraorde Scolecophidia. All have round (cycloid) scales that encircle the body. Their very small eyes are barely functional. These snakes generally occur near or in the colonies of the termites and ants on which they prey. They are often preyed upon by coral snakes (p. 157).

Anomalepis mexicanus — **Mexican Wormsnake**

JCN

18 cm (7.1 in)

Both head and body are uniformly brown. Note that 20 rows of scales encircle the body.

Helminthophis frontalis — **Pink-headed Wormsnake**

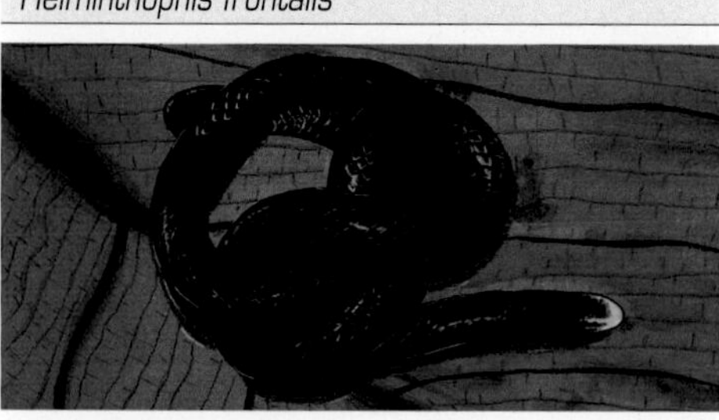

JCN

16 cm (6.3 in)

Brown on most of body but head and neck vary from yellowish to pink. Note that 20 rows of scales encircle the body.

Liotyphlops albirostris — **White-nosed Wormsnake**

Almost entirely brown; tip of the head is yellowish. Rostral and frontal scales are in contact with each other. Note that 20 rows of scales encircle the body.

22 cm (8.7 in)

Leptotyphlopidae (1 species)
Slender Blindsnakes

The slender blindsnake is the sole member of this family in Costa Rica. Family characteristics are nearly identical to those of Anomalepididae (p. 106).

Epictia [Leptotyphlops] goudotii [ater] **Slender Blindsnake**

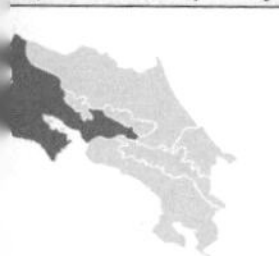

20 cm (8 in)

GF

The brown coloration on top of the body is a shade darker than the brown on either the sides of the body or the venter; head and tip of tail yellow to white. Note that 14 rows of scales encircle the body. Usually found in dry forests, under rocks and fallen branches.

Typhlopidae (1 species)
Blindsnakes

ne Costa Rican blindsnake is the sole member of this family in Costa Rica. Family characteristics are nearly identical to those of Anomalepididae (p. 106).

Typhlops costaricensis **Costa Rican Blindsnake**

36 cm (14 in)

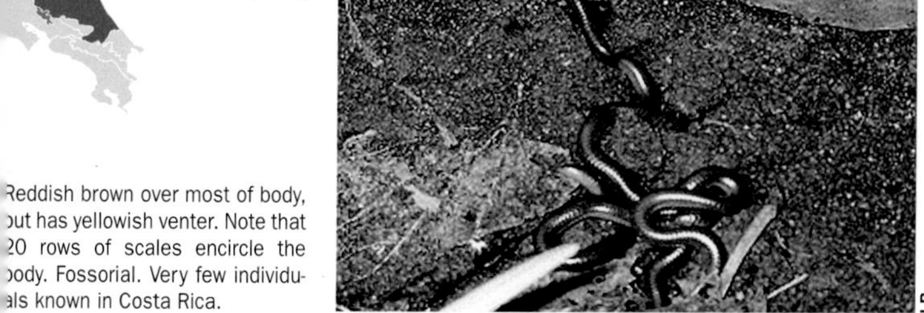

LB

Reddish brown over most of body, but has yellowish venter. Note that 20 rows of scales encircle the body. Fossorial. Very few individuals known in Costa Rica.

Loxocemidae (1 species)
Burrowing Python

The single species in this family shares many anatomical features with boas in the family Boidae (p. 109), including pelvic vestiges, extreme disengagement of lower jaw, and twin lungs. Has moderately small eyes. The burrowing python is primarily fossorial and nocturnal. This constrictor feeds on small mammals, lizards, and reptile eggs. Inhabits lowland dry forests, often near beaches.

Loxocemus bicolor — **Burrowing Python**

WB (RDJ)

1.4 m (4.5 ft)

Dorsal coloration is black to dark brown; venter varies from white to yellow. Snout is pointed and turned up.

Boidae (4 species)
Boas

Medium- to large-size constrictors. More than 38 rows of scales encircle the body. Head covered by small scales. Note absence of loreal pits that characterize snakes in the Viperidae family (p. 148); most species do have conspicuous labial pits. All species have vestigial pelvic spurs (larger in males) that hint at an ambulatory past. Boas are both terrestrial and arboreal and are mainly active at night.

Boa constrictor — Boa Constrictor

4.6 m (15 ft)

Dark- and light-brown blotches cover dorsum; note brown stripe down center of head; brown and tan bands encircle tip of tail. No labial pits. This strong constrictor has a robust body. The largest boa in Costa Rica.

RDJ

Corallus annulatus — Annulated Tree Boa

1.5 m (5 ft)

Dark transverse blotches overlay brown background. Also note large triangular head and large labial pits. Differs from *C. ruschenbergerii* (p. 110) in having nasal scales that are separated by two large internasal scales; lacks noticeable prefrontal scales. Its teeth (longer than those of most other boas) may allow it to more efficiently catch birds, its main prey. Arboreal.

RDJ

Corallus ruschenbergerii (hortulanus) **Garden Tree Boa**

RDJ

1.5 m (5 ft)

Dark transverse blotches overlay brown background. Also note large triangular head and large labial pits. Differs from *C. annulatus* (p. 109) in having nasal scales that are in contact with each other; has noticeable prefrontal scales. Its teeth (longer than those of most other boas) may allow it to more efficiently catch birds, its main prey. Arboreal.

Epricrates maurus (cenchria) **Rainbow Boa**

RDJ

1.8 m (6 ft)

Brown background coloration shows some iridescence; cylindrical body, labial pits present. Arboreal.

Ungaliophiidae (1 species)
Dwarf Boas

Apart from size, dwarf boas differ from other boids in having a single prefrontal scale (some boids lack a prefrontal scale, others have more than one) and in certain features of their bone structure. Like boids they have pelvic spurs that are visible in males. They are both arboreal and terrestrial. These secretive creatures are primarily nocturnal.

Ungaliophis panamensis — Dwarf Boa

65 cm (2.1 ft)

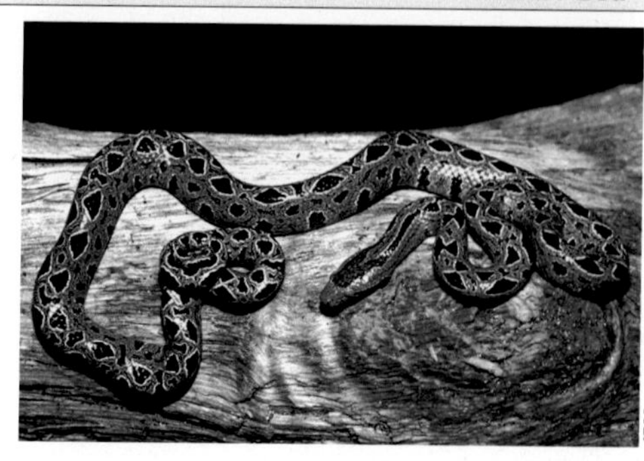

On dorsum, note dark geometric shapes (outlined by pale line) on light-brown background. Also note presence of one prefrontal scale.

Colubridae (105 species)
Colubrid Snakes

The largest family of snakes, with close to 65% of the species in the world, colubrids var greatly in color, habitat, and feeding habits. They range in size from a few inches to mor than 10 feet. In Costa Rica members of this family have large scales on the head excep for *Nothopsis rugosus* (p. 130), which has small scales. Most species are nonvenomous though some in Costa Rica have rear fangs with mild venom.

Amastridium veliferum — Ridge-nosed Snake

JPM

70 cm (2.3 ft)

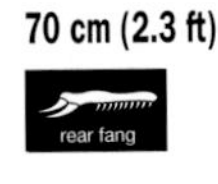

Dark gray to black over most of body; venter dark with white spots; head is rusty red. Snout ends in a well-defined ridge. Found mostly ir leaf litter, where it preys on frogs. Diurnal.

Chironius carinatus — Olive Keelback

RDJ

2.1 m (7 ft)

Dorsum is greenish brown; vente is cream yellow. Note large well-defined scales arranged i oblique rows. During the day preys mostly on frogs on the ground; at night climbs into tree to sleep. Aggressive. Similar t species of the genus *Leptophi* (pp. 126-128).

Chironius exoletus **Green Keelback**

1.6 m (5.2 ft)

RDJ

FMC

Dorsum is uniformly green; venter is greenish yellow. Note large, well-defined scales arranged in oblique rows. Forages on the ground during the day; at night climbs into vegetation to sleep. Aggressive. Similar to species of the genus *Leptophis* (pp. 126-128).

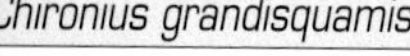

Chironius grandisquamis **Ebony Keelback**

2.4 m (8 ft)

RDJ

Dorsum is uniformly black; dark venter becomes paler toward front of body. Note large, well-defined scales arranged in oblique rows. Forages on the ground during the day; at night climbs into vegetation to sleep. Aggressive. Similar to species in the genus *Clelia* (p. 114).

Clelia clelia — Mussurana

RDJ (RDJ)
RDJ

2.4 m (8 ft)

A large snake with a round body. Shiny black dorsum contrasts with white venter. Juveniles of up to one meter (3.3 ft) in length are red overall, with black band on neck and snout and a white (to yellow) head. Known to feed on snakes, particularly pitvipers. Docile. Very similar to *C. scytalina* (below), though *Clelia clelia* occurs at lower elevations.

Clelia scytalina — Montane Mussurana

FMC

1.5 m (5 ft)

A large snake with a round body. Shiny black dorsum contrasts with white venter. Juveniles of up to one meter (3.3 ft) in length are red overall, with black band on neck and snout and a white (to yellow) head. Known to feed on snakes, particularly pitvipers. Docile. This is essentially a highland version of *C. clelia* (above).

Coniophanes bipunctatus **Twin-spotted Debris Snake**

76 cm (30 in)

JCN

Note diffuse black stripes on dark-brown background. Ventral scales with 2 black spots. Width of head equal to that of neck. Snout is tapered, ending in a point. Semiaquatic; feeds on fish and frogs.

Coniophanes fissidens **Brown Debris Snake**

80 cm (2.6 ft)

rear fang

AV

Note diffuse black stripes on dark-brown background. Ventral scales with many black spots. Width of head equal to that of neck. Snout is tapered, ending in a point. Bites can cause swelling and numbness in humans. Terrestrial.

Coniophanes piceivittis **Twin-striped Debris Snake**

61 cm (2 ft)

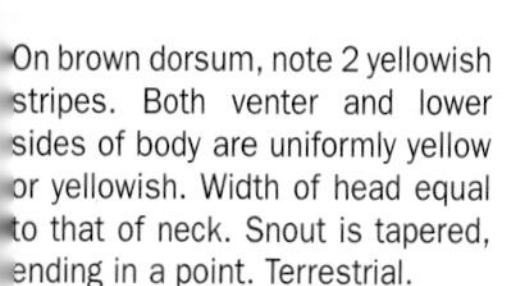

FMC

On brown dorsum, note 2 yellowish stripes. Both venter and lower sides of body are uniformly yellow or yellowish. Width of head equal to that of neck. Snout is tapered, ending in a point. Terrestrial.

Conophis lineatus — Road Guarder

RDJ

1.1 m (3.5 ft)

Alternating brown and yellow stripes run the length of the dorsum. Note brown labial scales. Width of head equal to that of neck. Snout is tapered, ending in a point. Bites can cause swelling, bleeding, and pain. Terrestrial.

Crisantophis nevermanni — Black Road Guarder

FMC

76 cm (2.5 ft)

Note 4 pale stripes that run the length of the dark-brown dorsum. Lower mandible shows conspicuous blotches. Ventral scales are cream color with black spots. Width of head equal to that of neck. Snout is tapered, ending in a point. Terrestrial.

Dendrophidion nuchale — Pink-tailed Forest Racer

AV

1.4 m (4.5 ft)

On front third of dorsum, color var ies from reddish brown to green rest of the dorsum is brown. Dorsa scales have notable keels; some individuals show a faint dorsal pat tern of irregular bands. Pale vente becomes reddish toward the tail Also note large eyes with rounc pupils, large head plates, and rela tively thin body. This fast-moving terrestrial snake feeds on frogs.

Dendrophidion paucicarinatum — **Cloud Forest Racer**

1.5 m (5 ft)

FMC

Dorsal scales (green with dark margins) form checkerboard pattern; venter is yellow. Also note large eyes with round pupils, large head plates, and relatively thin body. Anal plate usually divided. This fast-moving terrestrial snake feeds on frogs.

Dendrophidion percarinatum — **Brown Forest Racer**

1.2 m (3.9 ft)

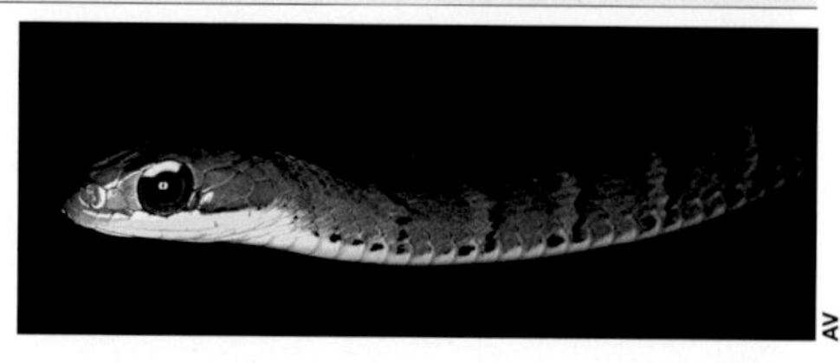

AV

On brown dorsum, note irregular rings that encircle body. Venter is yellow. Also note large eyes with round pupils and large head plates. Has a divided anal plate (*D. vinitor*, below, has a single anal plate). A fast-moving, thin snake. Terrestrial; feeds on frogs.

Dendrophidion vinitor — **Lowland Forest Racer**

1 m (3.1 ft)

WB

On brown dorsum note irregular bands; bands near the neck are more than 3 scales wide. Dorsal scales are notably keeled. Also note large eyes with round pupils, large head plates, and relatively thin body. Has a single anal plate. This fast-moving terrestrial snake feeds on frogs.

Dipsas articulata — Eastern Snail-eater

AV

70 cm (2.3 ft)

Note alternating pattern of black, cream, and red; red bands sometimes show just on dorsum. Very large eyes; relatively small, blunt head is well distinguished from neck and rest of body; body compressed laterally. Arboreal and nocturnal; feeds on snails and slugs. Similar to species of the genus *Sibon* (pp. 136-138). A coral snake mimic (p. 157).

Dipsas bicolor — Bicolor Snail-eater

RDJ

48 cm (1.6 ft)

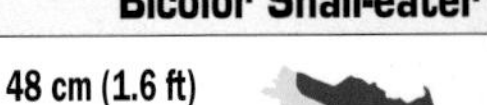

Wide black bands alternate with white ones; white bands with variable amount of orangish red on top. Very large eyes; relatively small, blunt head is well distinguished from neck and rest of body; body compressed laterally. Arboreal and nocturnal; feeds on snails and slugs. Similar to species of the genus *Sibon* (pp. 136-138). A coral snake mimic (p. 157).

Dipsas tenuissima — Southwestern Snail-eater

54 cm (21 in)

Wide black bands alternate with white ones (note varying amounts of reddish-brown spots on top of white bands). Very large eyes; relatively small, blunt head is well distinguished from neck and rest of body; body compressed laterally. Arboreal and nocturnal; feeds on snails and slugs. Similar to species of the genus *Sibon* (pp. 136-138).

Drymarchon melanurus (corais) **Black-tailed Cribo**

3 m (10 ft)

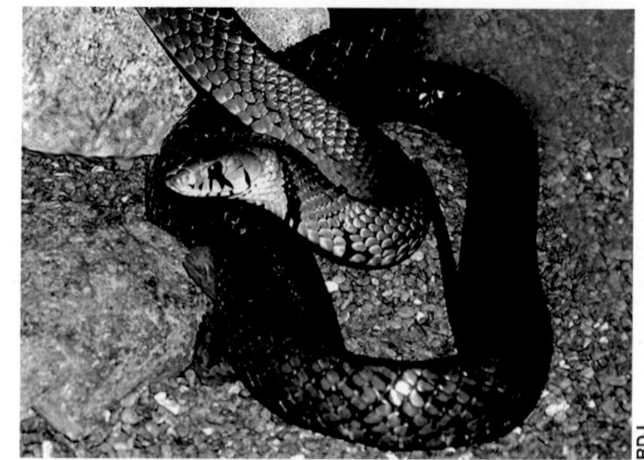

RDJ

Yellowish brown over most of body; also note black tail and black marks below eye and behind (and back from) neck. Aggressive.

Drymobius margaritiferus **Speckled Racer**

1.2 m (4 ft)

Dorsal scales (yellowish green with dark margins) create a checkerboard pattern; yellow venter. Has large eyes with round pupils, large head plates, and a relatively thin body. This fast-moving terrestrial snake feeds on frogs.

WB

Drymobius melanotropis **Green Racer**

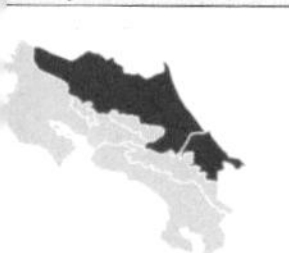

1.4 m (4.5 ft)

rear fang

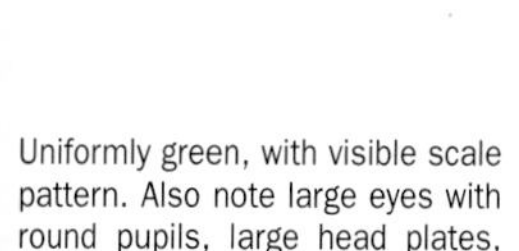

Uniformly green, with visible scale pattern. Also note large eyes with round pupils, large head plates, and a relatively thin body.

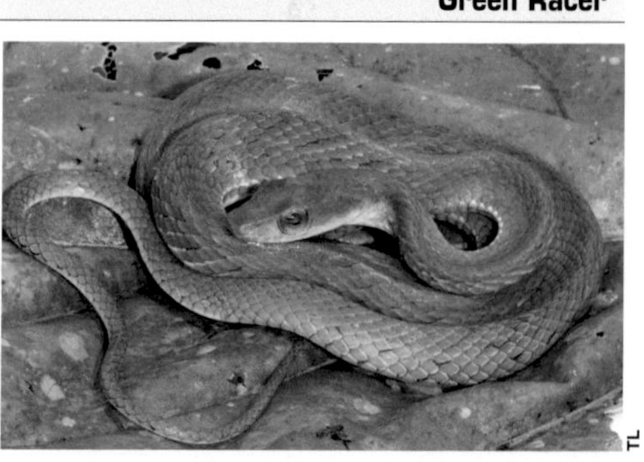

TL

Drymobius rhombifer — Blotched Racer

DH

1.2 m (3.9 ft)

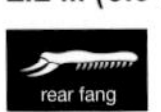

A series of rhomboids (each with light-brown center and black borders) appears along the body. Also has large eyes with round pupils, large head plates, and a relatively thin body. This fast-moving terrestrial snake feeds on frogs.

Enuliophis (Enulius) sclateri — White-headed Snake

RDJ

55 cm (22 in)

Dorsum is uniformly black; head is white except for black tip of snout. Slender body. Mostly found under leaf litter. Feeds on eggs of small reptiles.

Enulius flavitorques — Collared Snake

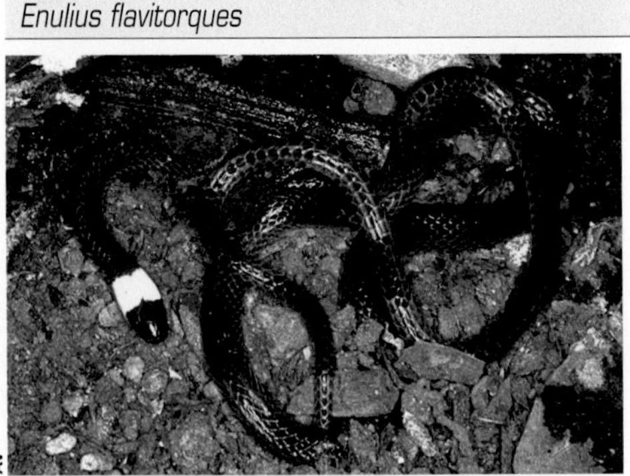

AV

51 cm (20 in)

Dorsum is uniformly brown; note white neck band. Slender. Generally found under leaf litter. Feeds on eggs of small reptiles.

Erythrolamprus bizonus — Brilliant Ringed Snake

1.1 m (3.5 ft)

A coral snake mimic (p. 157); red, black, and white (or yellow) rings encircle the body. Red rings are 3 times wider than rings of another color; white rings always fall between black rings, never touching red. Note small white scales on black head. Bites can cause pain and swelling in humans.

Erythrolamprus mimus — Ringed Snake

70 cm (2.3 ft)

rear fang

A coral snake mimic (p. 157); red, black, and white (or yellow) rings encircle the body. Red rings are 3 times wider than rings of another color; black rings (surrounded by white) never touch red rings. Note small white scales and white ring on black head. Bites can cause pain and swelling in humans.

Geophis brachycephalus — Gray Earth Snake

50 cm (20 in)

Color of dorsum varies from shiny black to dark brown, with some individuals showing red blotches. Some juveniles have a white neck band. Snout is tapered, ending in a point. Mostly found under decaying logs or leaf litter. Common. (The *Geophis* species that occur in Costa Rica are very similar to one another and are extremely difficult to distinguish in the field.)

Geophis downsi — Down's Earth Snake

Dorsum is shiny dark brown. Some individuals have white neck band. Snout is tapered, ending in a point. Mostly found under decaying logs or under leaf litter. Rare; known only from Las Cruces (near border with Panama), at 4,000 ft.

25 cm (10 in)

Geophis godmani — Yellow-bellied Earth Snake

RDJ

40 cm (16 in)

On dorsum, note shiny brown scales and yellow interspaces; venter yellow. Snout is tapered, ending in a point. Found under decaying logs or under leaf litter. Common.

Geophis hoffmanni — Hoffman's Earth Snake

FMC

30 cm (12 in)

Dorsal coloration varies from shiny black to dark brown. Some juveniles have a white neck band. Mostly found under decaying logs or leaf litter. Common.

Geophis ruthveni — Ruthven's Earth Snake

Dorsal coloration varies from shiny black to dark brown. Snout is tapered, ending in a point. Mostly found under decaying logs or leaf litter. Rare.

26 cm (10 in)

Geophis talamancae — Talamanca Earth Snake

Dorsal coloration varies from shiny black to dark brown. Snout is tapered, ending in a point. Mostly found under decaying logs and leaf litter. Rare; known only from Cerro Pando (near border with Panama), at 6,200 ft.

22 cm (9 in)

Geophis zeledoni — Zeledon's Earth Snake

40 cm (16 in)

Dorsal coloration varies from shiny black to dark brown. Snout is tapered, ending in a point. Mostly found under decaying logs and leaf litter. Rare; known only from the area between Poás and Barva volcanos.

Hydromorphus concolor — Tropical Seep Snake

80 cm (2.6 ft)

Dorsum is uniformly brown, with smooth scales. Has upward pointing nostrils. Cylindrical body. Aquatic, feeds on freshwater shrimp.

Imantodes cenchoa — Blunt-headed Tree Snake

1.3 m (4.3 ft)

Light-brown snake with large, angular, vertebral blotches along body, which is extremely thin and vertically compressed. Enlarged head and eyes. Large vertebral scales are 3 or more times wider than lateral scales. Arboreal.

Imantodes gemnistratus **Forest Blunt-headed Tree Snake**

Light-brown snake with large, irregular vertebral bands along body that continue down to venter. Extremely thin, vertically compressed body. Enlarged head and eyes. Large vertebral scales are wider (to varying degree) than lateral scales. Arboreal.

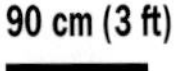

90 cm (3 ft)

Imantodes inornatus **Plain Blunt-headed Tree Snake**

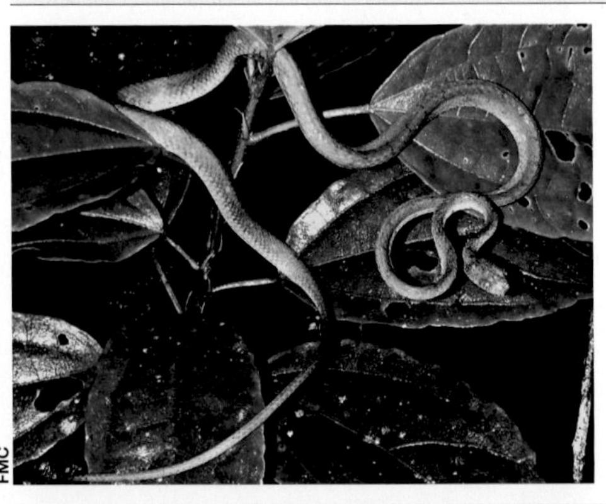

FMC

1.1 m (3.5 ft)

Dorsum is uniformly brown; also note dorsal scales with small black specks. Extremely thin, compressed body. Enlarged head and eyes. Arboreal.

Lampropeltis triangulum **Costa Rican Milk Snake**

RDJ

RDJ

1.5 m (5 ft)

A coral snake mimic (p. 157). Red, black, and white rings encircle the body. Red rings are twice as wide as black and white rings; note that white rings always fall between 2 black rings, never touching a red ring. Head is black with a white ring near neck. Uniformly black individuals tend to be more common at high elevations. Constrictor.

Leptodeira annulata — **Annulated Cat-eyed Snake**

1 m (3.1 ft)

RDJ

On dorsum, note dark-brown round markings on light-brown background. On large head, center stripe meets first dorsal marking. Large eyes with vertical pupils. Smooth scales. Mostly arboreal.

Leptodeira nigrofasciata — **Black-banded Cat-eyed Snake**

60 cm (2 ft)

rear fang

RDJ

Dorsum shows wide black bands on a creamy background. Venter cream. Large eyes with vertical pupils.

Leptodeira rubricata — **Barred Cat-eyed Snake**

70 cm (2.3 ft)

AV

On dorsum, brown bands alternate with orange bands; lower flanks and venter are light brown. Large eyes with vertical pupils. Generally found on vegetation near or in mangroves.

Leptodeira septentrionalis — Northern Cat-eyed Snake

RDJ

1 m (3.1 ft)

On dorsum, note dark-brown roundish markings on a light-brown background. Large eyes with vertical pupils. Smooth scales. Arboreal.

Leptodrymus pulcherrimus — Green-headed Racer

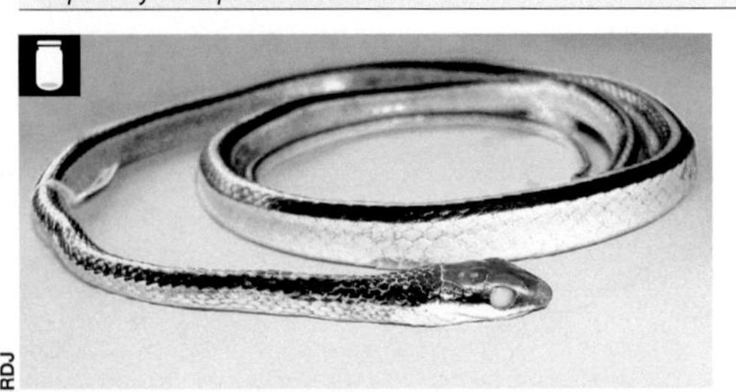
RDJ

1.5 m (5 ft)

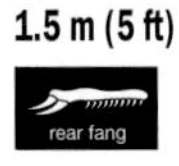

Green head. Two black stripes from eyes to tail on a cream background. Terrestrial. Diurnal.

Leptophis ahaetulla — Green Parrot Snake

RDJ (RDJ)

2 m (6.5 ft)

rear fang

Green dorsum; light-green venter. On large head, note white mandible; eyes have yellow iris crossed in the middle by a black stripe. Loreal scale absent. Mostly arboreal. Similar to several species in the genus *Chironius* (pp. 112-113).

Leptophis depressirostris **Eastern Parrot Snake**

1.5 m (4.9 ft)

Green dorsum; light-green venter. On large head, note yellow mandible; eyes have yellow iris crossed in the middle by a black stripe. Loreal scale present (absent in *L. ahaetulla*, p. 126). Mostly arboreal. Similar to several species in the genus *Chironius* (pp. 112-113).

Leptophis mexicanus **Mexican Parrot Snake**

1.3 m (4.3 ft)

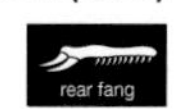

Alternating gold and green stripes form zigzag pattern on dorsum; light-cream venter. On large head, note white mandible; above mandible, a black stripe passes through yellow iris and terminates past neck. Loreal scale present. Mostly arboreal. Similar to several species in the genus *Chironius* (pp. 112-113).

Leptophis nebulosus **Bronze-striped Parrot Snake**

60 cm (2 ft)

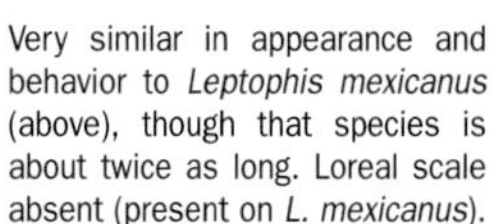

Very similar in appearance and behavior to *Leptophis mexicanus* (above), though that species is about twice as long. Loreal scale absent (present on *L. mexicanus*). Mostly arboreal.

Leptophis riveti — **Turquoise Parrot Snake**

AV

1 m (3.1 ft)

On dorsum, note alternating green (with turquoise hues) and golden (or dark-brown) bands; light-green venter. On large head, note yellowish mandible; black stripe passes through yellow iris. Mostly arboreal.

Liophis epinephalus — **Fire-bellied Snake**

FMC (AV)

80 cm (2.6 ft)

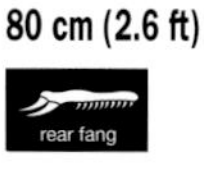

On some individuals, dorsum is uniformly brown; on others, note alternating brown and reddish bands that become more conspicuous when the snake expands its body. Red-orange venter. Flattens its body when threatened.

Masticophis (Coluber) mentovarius — **Dry Forest Whipsnake**

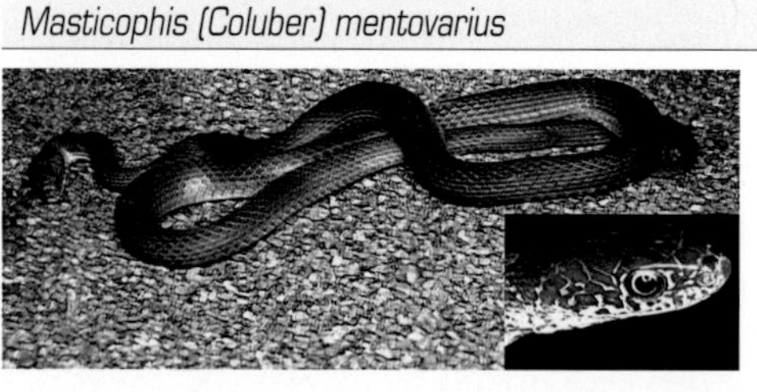

RDJ (AV)

2 m (6.5 ft)

Color on dorsum varies from light green to brown depending on the individual. Spots on side of head are characteristic. A fast-moving terrestrial snake.

Mastigodryas (Dryadophis) melanolomus **Salmon-bellied Racer**

1.4 m (4.5 ft)

RDJ (AV, AV)

On adults, 2 dark stripes run along each side of body; on venter, salmon color changes to yellowish toward head. On juveniles, note brown blotches on sides.

Ninia celata **Black Wood Snake**

45 cm (18 in)

Uniformly black dorsum; pale venter. Some individuals have a white nuchal band. Small head and tiny eyes. Secretive, occurring mostly under leaf litter. Flattens its body when threatened.

Ninia maculata **Spotted Wood Snake**

36 cm (14 in)

AV

RDJ (RDJ)

MRV (MRV)

Extremely variable in coloration, with background ranging from brown to red; many individuals have narrow crossbands on dorsum. Note checkered venter. Small head and tiny eyes. Secretive, occurring mostly under leaf litter. Flattens its body when threatened.

Ninia psephota **Dark Wood Snake**

FMC (RDJ)

50 cm (20 in)

Scales on dorsum are gray but note pale interscale areas; venter is checkered. Also note small head and eyes. Secretive, occurring mostly under leaf litter. Flattens its body when threatened.

Ninia sebae **Ring-necked Coffee Snake**

RDJ

40 cm (16 in)

Dorsum and sides show a beautiful coppery-red pattern; behind black head, note yellow and black neck bands. Small head and eyes. Secretive, occurring mostly under leaf litter. Flattens its body when threatened.

Nothopsis rugosus **Diamond Water Snake**

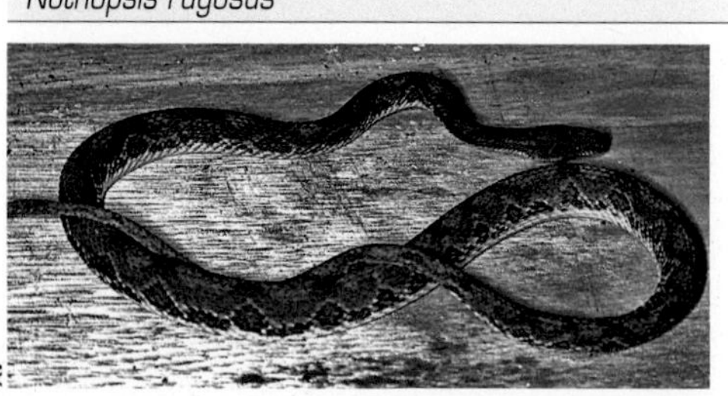

AV

44 cm (17 in)

The sole species in its genus. Note dark triangular shapes on light-brown body. Head covered with many small scales.

Oxybelis aeneus — Brown Vine Snake

1.8 m (6 ft)

Very slender snake; head (very elongated and pointed) is brown above and white on underside. Dorsum is uniformly brown, venter is white. Often found on vegetation. When threatened, opens mouth wide to display dark-purple lining.

RDJ (RDJ)

Oxybelis brevirostris — Short-nosed Vine Snake

1.2 m (4 ft)

Very slender snake; head is elongated and pointed. Shows uniform green coloration. Mostly found on vegetation.

RDJ (RDJ)

Oxybelis fulgidus — Green Vine Snake

2.1 m (7 ft)

rear fang

Very slender snake; head is very elongated and pointed. Green overall, with a yellow stripe down each side of body. Mostly found on vegetation. Bites can cause pain and swelling.

RDJ (RDJ)

Oxyrhopus petola (petolarius) — Bush Racer

RDJ (RDJ, RDJ)

1.2 m (3.9 ft)

Black and red bands somewhat resemble rings found on the Many-banded Coral Snake (*Micrurus multifaciatus*, p. 159), though this species has a larger head, larger eyes, and a pearly white venter.

Pliocercus (Urotheca) euryzonus — Halloween Snake

AV

80 cm (2.6 ft)

Resembles a coral snake (p. 157). Shows pattern of alternating black and orange (or red) bands; the orange bands become white on the lower portion of the sides of the body. Tip of snout is black. Small head and eyes. Secretive, found mostly under leaf litter.

Pseustes poecilonotus — Bird Snake

RDJ

2.4 m (8 ft)

Adults vary greatly in pattern and coloration, with blotches of yellow, brown, green, and black. On juveniles, note orange bands (with dark edges) on body, and dark stripes on sides of head. Diurnal; semiarboreal; and often aggressive.

Rhadinaea calligaster — **Green Litter Snake**

52 cm (20 in)

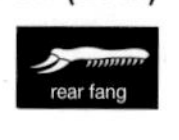

FMC

Along length of body run alternating green-brown and yellow stripes; venter yellow overall but note dark scale borders. Small head and eyes. Secretive, occurring mostly under leaf litter.

Rhadinaea decorata — **Pink-bellied Litter Snake**

51 cm (20 in)

AV

Brown over most of dorsum, with a pale stripe running the length of the body along each side; venter is bright orange. Note small head (with bright yellow spot behind the head) and small eyes. Secretive, generally found under leaf litter.

Rhadinaea pulveriventris — **Dusty-bellied Litter Snake**

50 cm (20 in)

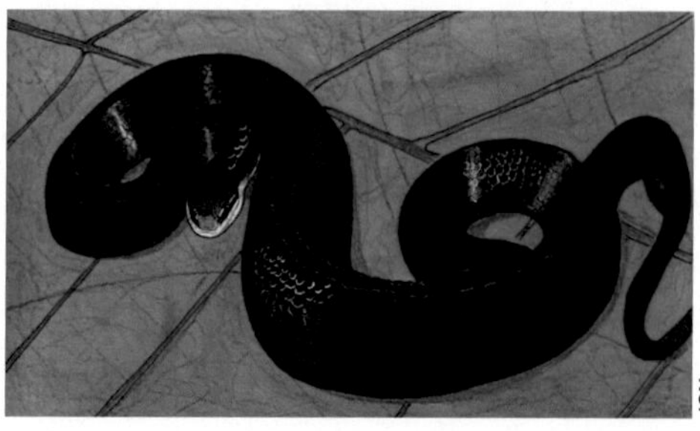
JCN

On deep-brown dorsum, note dark stripe that extends from eyes along one-third the length of the body. Venter pale with multiple minute spots. Also note small head and small eyes. Secretive, occurring mostly under leaf litter.

Rhadinella (Rhadinaea) godmani — Godman's Brown Snake

TL

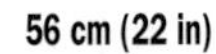

56 cm (22 in)

On light-brown background; note multiple dark stripes on dorsum running along the length of the body; yellow venter. Small head, small eyes. Secretive, found mostly under leaf litter.

Rhadinella (Rhadinaea) serperaster — Striped Litter Snake

RDJ

45 cm (18 in)

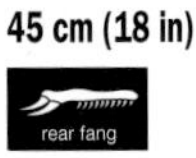

On light-brown background, note multiple light stripes running the length of the dorsum and sides; white venter. Small head, small eyes. Secretive, found mostly under leaf litter.

Rhinobothryum bovallii — Costa Rican Tree Snake

RDJ

1.7 m (5.6 ft)

Resembles a coral snake (p. 157). Narrow white rings fall between wide red and black rings. Black head scales are bordered by white. Arboreal.

Scaphiodontophis annulatus (venustissimus) **False Coral Snake**

92 cm (3 ft)

RDJ

Very closely resembles a coral snake (p. 157). Narrow yellow bands fall between wide red and black bands. On some individuals, posterior section of body is brown with black specks. White venter. Terrestrial and very fast moving.

Scolecophis atrocinctus **Harlequin Snake**

45 cm (18 in)

AV

Resembles a coral snake (p. 157). Black bands alternate with white bands (the white bands become orange or red on the dorsum). Tip of snout white.

Senticolis (Elaphe) triaspis **Green Rat Snake**

1.2 m (4 ft)

RDJ (FMC)

Color of dorsum varies from individual to individual, from brown to yellowish green. Cream-colored venter. Mostly terrestrial but also climbs vegetation. Constrictor.

Sibon annulatus — Ringed Snail-eater

AV

56 cm (22 in)

Complex coloration pattern; note brown and white bands (white bands have variable amounts of mossy green on dorsum). Very large eyes in a relatively small, blunt head; head clearly distinguished from neck and rest of body. Body compressed laterally. Arboreal and nocturnal. Feeds on snails and slugs. Resembles species of the genus *Dipsas* (p. 118).

Sibon anthracops — Red-ringed Snail-eater

RDJ

56 cm (22 in)

Wide black bands alternate with narrow white bands; on dorsum, white bands show varying amounts of orangish red. Body is laterally compressed. Has very large eyes; blunt head, though relatively small, is clearly distinguishable from neck. Arboreal and nocturnal. Feeds on snails and slugs. Resembles species of the genus *Dipsas* (p. 118). A coral snake mimic (p. 157).

Sibon argus — Blotched Snail-eater

70 cm (2.3 ft)

Similar in appearance and behavior to *Sibon longifrenis* (p. 138). On dorsum, note semicircular brown blotches (margined by black) on a lichen-green background. Body is laterally compressed. Has very large eyes; blunt head, though relatively small, is clearly distinguishable from neck. Arboreal and nocturnal. Feeds on snails and slugs. Resembles species of the genus *Dipsas* (p. 118).

Sibon dimidiatus **Banded Snail-eater**

66 cm (2.2 ft)

FMC

RDJ

Highly variable in appearance. Note brown, rust, or red bands; pale venter extends to lateral area between bands, becoming (on most individuals) tan or gray on dorsum. Body is laterally compressed. Very large eyes; blunt head, though relatively small, is clearly distinguishable from neck. Arboreal and nocturnal. Feeds on snails and slugs. Resembles species of the genus *Dipsas* (p. 118).

Sibon lamari **Costa Rican Snail-eater**

55 cm (22 in)

RVL

Similar in appearance to *Sibon longifrenis* (p. 138) but with wider head. Very similar in behavior. Known only from Guayacán de Siquirres. Resembles species in the genus *Dipsas* (p. 118).

Sibon longifrenis **Lichen-colored Snail-eater**

DH

64 cm (2.1 ft)

On dorsum, note semicircular brown blotches (margined by black) on a lichen-green background. Body is laterally compressed. Has very large eyes; blunt head, though relatively small, is clearly distinguishable from neck. Arboreal and nocturnal. Feeds on snails and slugs. Resembles species in the genus *Dipsas* (p. 118).

Sibon nebulatus **Cloudy Snail-eater**

RDJ

84 cm (2.8 ft)

Dark-brown and gray bands form an irregular pattern. Body is laterally compressed. Very large eyes; blunt head, though relatively small, is clearly distinguishable from neck. Arboreal and nocturnal. Feeds on snails and slugs. Resembles species in the genus *Dipsas* (p. 118).

Siphlophis (Tripanurgos) compressus **Red-eyed Tree Snake**

RDJ (RDJ)

1.2 m (4 ft)

Orange on front of head fades to yellow toward the neck; note wide black nuchal band; dorsum is pinkish red, intersected by a few short, black crossbands. Venter creamy white. Arboreal.

Spilotes pullatus — Tiger Rat Snake

2.4 m (8 ft)

RDJ

RDJ

On dorsum, black background interrupted by oblique yellow lines; on venter, yellow background interrupted by black lines. Head with transverse black and yellow lines. Aggressive.

Stenorrhina degenhardtii — Brown Scorpion-eater

82 cm (2.7 ft)

On dorsum, note dark-brown irregular blotches on a light-brown background. The fusion of internasal scales and prenasal scales results in fewer scales between the nostrils and creates a shovel-like appearance. Round pupils. Primarily feeds on scorpions and tarantulas.

JPM

Stenorrhina freminvillii **Red Scorpion-eater**

WH
FMC

82 cm (2.7 ft)

Color of dorsum varies from individual to individual, ranging from orangish brown to pinkish red. The fusion of internasal scales and prenasal scales results in fewer scales between the nostrils and creates a shovel-like appearance. Round pupils. Primarily feeds on scorpions and tarantulas.

Tantilla alticola **Brown Crowned Snake**

TL

32 cm (13 in)

rear fang

Has brown dorsum and nuchal band; red venter. Neck indistinguishable from body. Small eyes. Secretive, found mostly under leaf litter.

Tantilla armillata — Black-necked Crowned Snake

49 cm (19 in)

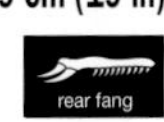

JCN

Head shows black cap and white spots around the mouth; on dark-brown dorsum, note multiple faint stripes along length of body; cream-colored venter. Neck indistinguishable from body. Small eyes. Secretive, found mostly under leaf litter.

Tantilla reticulata — Lined Crowned Snake

32 cm (13 in)

GF

Down center of dorsum, note pale wide stripe; on dorsum and sides, also note multiple light and dark stripes along the length of the body. Yellow venter. Neck indistinguishable from body. Small eyes. Secretive, found mostly under leaf litter.

Tantilla ruficeps (melanocephala) — Striped Crowned Snake

50 cm (20 in)

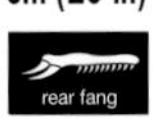

FMC (FMC)

Area around mouth is creamy white (except under eyes). On dorsum and sides, note brown background with multiple stripes that run the length of the body. Red venter. Neck indistinguishable from body. Small eyes. Secretive, found mostly under leaf litter.

Tantilla schistosa — Red-tailed Crowned Snake

FMC (FMC, FMC)

35 cm (14 in)

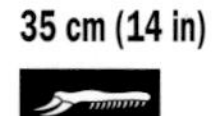

White (to orange) nuchal band; dorsum is uniformly dark brown; red venter. Neck indistinguishable from body. Small eyes. Secretive, found mostly under leaf litter.

Tantilla supracincta — Coral Crowned Snake

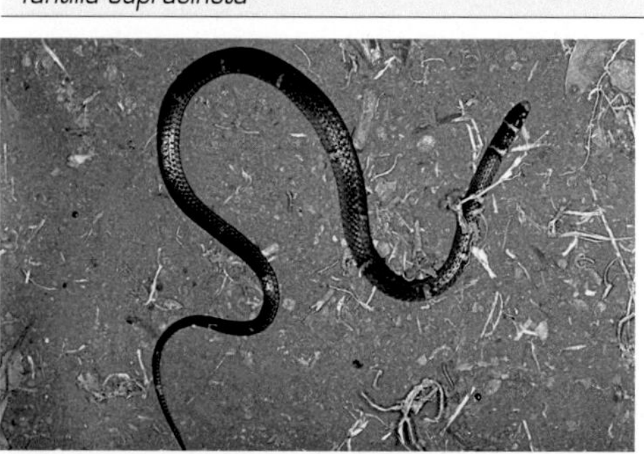

AV

66 cm (2.2 ft)

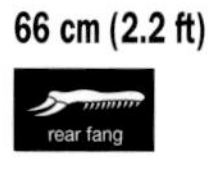

Somewhat resembles a coral snake (p. 157). On red background, note yellow bands between 2 black bands. Red venter. Neck indistinguishable from body. Small eyes. Secretive, found mostly under leaf litter.

Tantilla vermiformis — Plain-bellied Crowned Snake

On light-brown dorsum, note central dark stripe running along length of body. Cream-colored venter. Neck indistinguishable from body. Small eyes. Secretive, found mostly under leaf litter.

16 cm (6 in)

Thamnophis marcianus — Checkered Garter Snake

1.1 m (3.5 ft)

RDJ

On dorsum, black and light-brown shapes form a checkerboard pattern. Often found near water. Flattens neck when threatened. Uncommon in Costa Rica.

Thamnophis proximus — Ribbon Snake

1.2 m (4 ft)

RDJ

Dorsum is brown; on each side, a yellow stripe runs along the first half of the body; white venter. Also note yellow around the mouth. Often found near water. Holds mouth agape when threatened.

Tretanorhinus nigroluteus — Orange-bellied Swamp Snake

85 cm (2.8 ft)

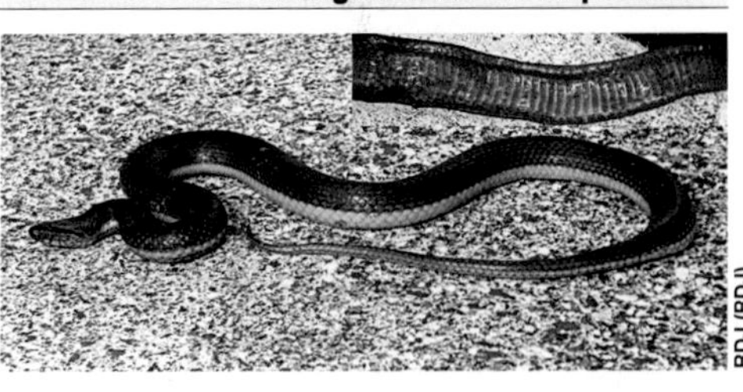

RDJ (RDJ)

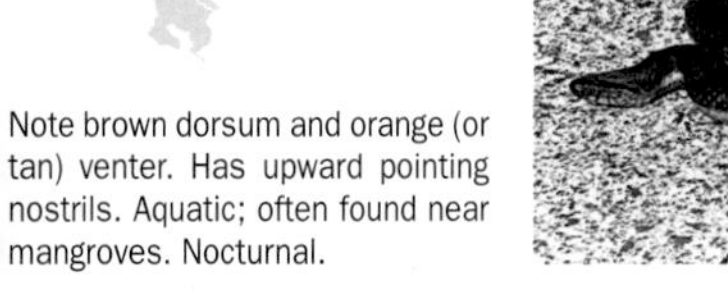

Note brown dorsum and orange (or tan) venter. Has upward pointing nostrils. Aquatic; often found near mangroves. Nocturnal.

Trimetopon gracile — Slender Dwarf Snake

RDJ

30 cm (12 in)

Note combination of faint nuchal spots and alternating clear and dark stripes along length of body. Cream-yellow venter. Neck indistinguishable from body. Small eyes. Secretive, found mostly under leaf litter.

Trimetopon pliolepis — Faded Dwarf Snake

FMC (FMC)

28 cm (11 in)

Very similar to *Trimetopon gracile* (above) in both appearance and behavior.

Trimetopon simile — White-bellied Dwarf Snake

Note nuchal band; uniformly brown dorsum; cream-white venter. Neck indistinguishable from body. Small eyes. Secretive, found mostly under leaf litter.

18 cm (7 in)

Trimetopon slevini — Slevin's Dwarf Snake

Note combination of nuchal band, brown stripes on dorsum, and red venter. Neck indistinguishable from body. Small eyes. Secretive, found mostly under leaf litter.

30 cm (12 in)

Trimetopon viquezi — Viquez's Dwarf Snake

Note absence of nuchal band (both *T. simile* and *T. slevini*, above, have a nuchal band). Also note dark stripes on dorsum and red venter. Neck indistinguishable from body. Small eyes. Secretive, found mostly under leaf litter.

25 cm (10 in)

Trimorphodon quadruplex (biscutatus) **Lyre Snake**

1.8 m (6 ft)

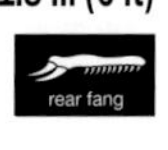

On dorsum, note dark, butterfly-like blotches on a light-brown background. Head shows lyre-shaped blotch. Though generally terrestrial, this snake is an adept climber. Diverse diet includes lizards, birds, and bats.

Tropidodipsas sartorii **Northern Snail-eater**

80 cm (32 in)

Shows pattern of black and white bands. Has very large eyes. Small blunt head is clearly distinguishable from neck and rest of body. Body compressed laterally. Arboreal and nocturnal. Feeds on snails and slugs. Similar to species in the genera *Dipsas* (p. 118) and *Sibon* (pp. 136-138).

Urotheca decipiens **Long-tailed Litter Snake**

61 cm (2 ft)

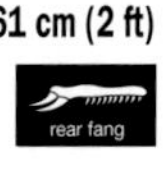

Note bright yellow nuchal band. Dorsum varies from dark brown to black depending on individual. A pale stripe runs along each side of the body. Neck indistinguishable from body. Small eyes. Secretive, found mostly under leaf litter.

Urotheca fulviceps — Tawny-headed Litter Snake

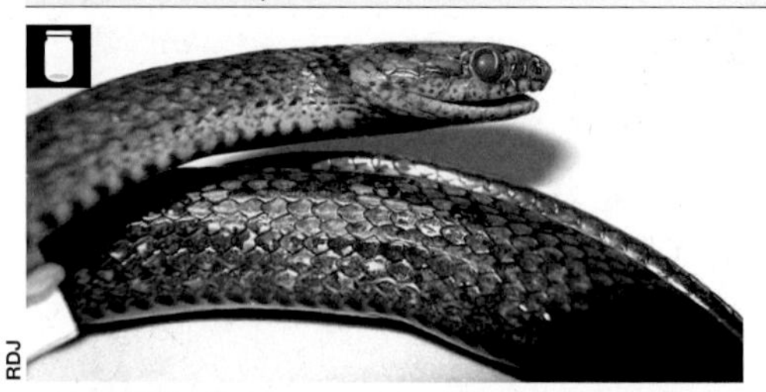
RDJ

65 cm (2.1 ft)

Reddish head cap; uniformly brown dorsum and white venter. Neck indistinguishable from body. Small eyes. Secretive, found mostly under leaf litter.

Urotheca guentheri — Orange-bellied Litter Snake

FMC

68 cm (2.2 in)

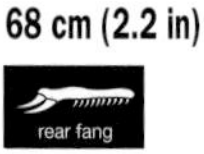

Note series of bright yellow spots behind the eye. Dorsum brown; on each side, two pale stripes run the length of the body. Bright orange venter. Neck indistinguishable from body. Small eyes. Secretive, found mostly under leaf litter.

Urotheca myersi — Myers' Litter Snake

Dark nuchal band; uniformly brown dorsum; and yellow venter. Neck indistinguishable from body. Small eyes. Secretive, found mostly under leaf litter.

35 cm (14 in)

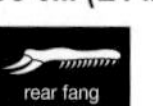

Urotheca pachyura — Thick-tailed Litter Snake

TL

68 cm (2.2 ft)

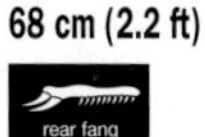

Note reddish-brown head. Dorsum is dark brown; on each side, a faint stripe runs the length of the body. White venter. Neck indistinguishable from body. Small eyes. Secretive, found mostly under leaf litter.

80 cm (2.6 ft)

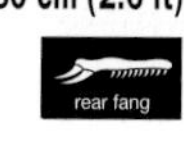

Note dark rhomboids on dorsum, a pattern similar to that of the venomous *Bothrops asper* (p. 152). Distinguished from that species by round pupils and brown scale interspaces around the mouth. Flattens body when threatened.

Viperidae (16 species)
Pitvipers

The snakes in this family are the most specialized predators among all snakes. The long, hollowed, retractable fangs stab and inject lethal venom into prey animals. Larg mouths—and the ability to disengage the mandibles for easier swallowing—make it po sible for them to consume relatively large prey. Heat sensors in pits between eyes an nostrils allow them to detect passing prey. Although pitvipers often spend many days i the same spot passively waiting for prey, most will react aggressively if threatened. Th majority of species are terrestrial, a few are arboreal. Terrestrial species are often foun coiled, with their head resting on the top of the coil. Their coloration tends to mimic the surroundings, especially when coiled. These snakes are responsible for more than 90% the fatal snakebites in Costa Rica.

Agkistrodon bilineatus **Cantil**

RDJ

1.2 m (4 ft)

On each side of face, 2 white lines run from nostril toward back of head. Also note vertical pupils and loreal pit (between nostril and eye). On dorsum, dark- and light brown blotches are divided by very dark irregular lines; dorsum shows highly keeled scales. Has a robust body. Triangular head clearly distinguishable from neck Tail is short and distinct from res of body. Often found near water.

Atropoides mexicanus (nummifer) **Jumping Pitviper**

76 cm (2.5 ft)

Note dark stripe behind each eye. Also note vertical pupils and loreal pit (between nostril and eye). On dorsum, dark-brown rhomboids overlay a light-brown background; dorsum has extremely keeled scales. Triangular head clearly distinguishable from neck. Body is robust, thick, and short. Tail is short and distinct from rest of body. Rostral and nasal scales do not come into contact with each other (see *Atropoides picadoi*, below).

RDJ

Atropoides picadoi **Picado's Pitviper**

91 cm (3 ft)

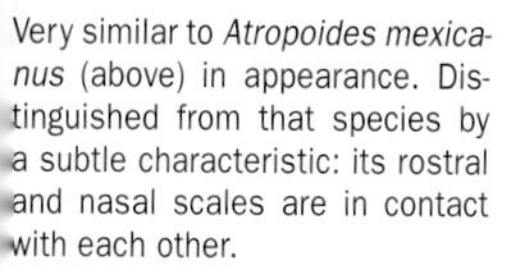

Very similar to *Atropoides mexicanus* (above) in appearance. Distinguished from that species by a subtle characteristic: its rostral and nasal scales are in contact with each other.

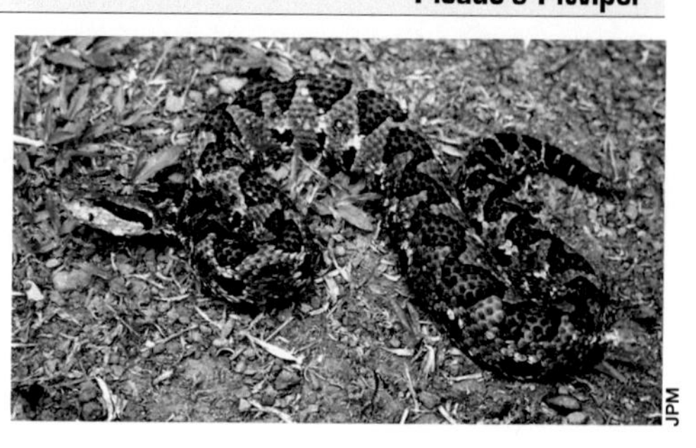

JPM

Bothriechis lateralis **Green Palm Pitviper**

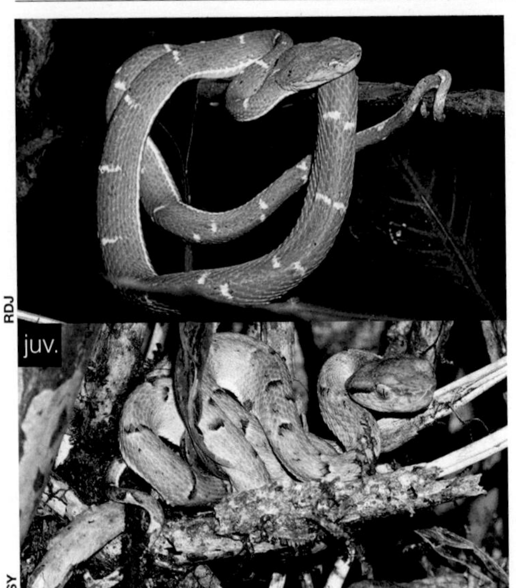

RDJ

juv.

SY

91 cm (3 ft)

viper fang

Dorsum is leaf-green color, with narrow yellow transverse lines along body; dorsum has extremely keeled scales. Also note vertical pupils and loreal pit (between nostril and eye). Triangular head clearly distinguishable from neck. Robust body. Tail is short and distinct from rest of body. Arboreal. Juveniles are brown to greenish brown.

Bothriechis nigroviridis **Black & Green Palm Pitviper**

MRV

76 cm (2.5 ft)

viper fang

Note yellow face and mossy-green pattern on dorsum; dorsum has extremely keeled scales. Also note vertical pupils and loreal pit (between nostril and eye). Triangular head clearly distinguishable from neck. Robust body. Tail is short and distinct from rest of body. Arboreal.

Bothriechis schlegelii **Eyelash Pitviper**

76 cm (2.5 ft)

FMC

RDJ

Elevated scales on dorsal margin of eye form 2 eyelash-like projections; also note vertical pupils. Color of dorsum varies greatly; even among individuals of the same brood, some may be bright yellow, while others have a lichen-like pattern in tones of brown, gray, and green. Dorsum has extremely keeled scales. Has loreal pits (between each nostril and eye). Triangular head clearly distinguishable from neck. Robust body. Tail is short and distinct from rest of body. Arboreal.

Bothriechis supraciliaris **Blotched Eyelash Pitviper**

60 cm (2 ft)

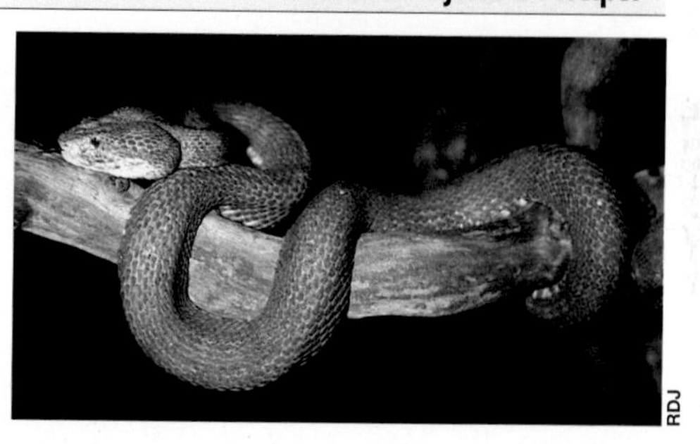

RDJ

Elevated scales on dorsal margin of eye form two eyelash-like projections; also note vertical pupils. Dorsal coloration varies greatly in this species. Even individuals of the same brood can be born with various tones of brown mixed with gray or green; dorsum has extremely keeled scales. Has loreal pits (between each nostril and eye). Triangular head clearly distinguishable from neck. Robust body. Tail is short and distinct from rest of body. Mostly arboreal.

2.3 m (7.5 ft)

viper fang

Note dark postocular stripe; also note vertical pupils. Coloration and pattern are somewhat variable. On dorsum, light-brown rhomboids appear on a dark-brown background; dorsum has extremely keeled scales. Has loreal pits (between each nostril and eye). Triangular head clearly distinguishable from neck. Robust body. Tail is short and distinct from rest of body. Terrestrial, though smaller individuals sometimes climb bushes.

Cerrophidion godmani **Godman's Pitviper**

76 cm (2.5 ft)

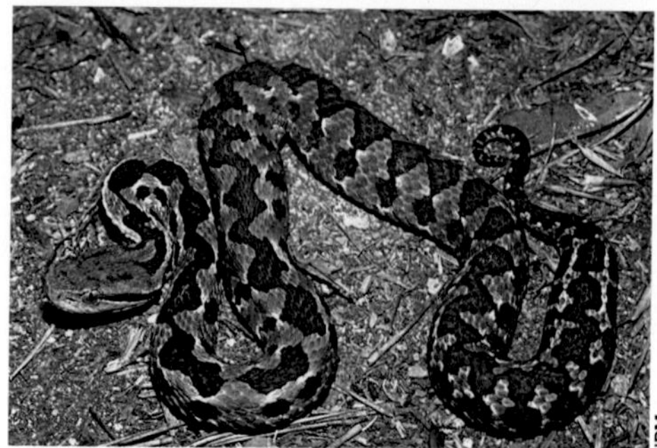
JPM

Note dark postocular stripe; also note vertical pupils. On light-brown dorsum appear dark-brown irregular rhomboids that are fused; dorsum has extremely keeled scales. Has loreal pits (between each nostril and eye). Triangular head clearly distinguishable from neck. Robust body. Tail is short and distinct from rest of body. Terrestrial.

Crotalus simus (durissus) **Central American Neotropical Rattlesnake**

1.7 m (5.5 ft)

RDJ

From each eye, 2 wide brown stripes run well past the neck; also note vertical pupils. On light-brown dorsum, note dark-brown rhomboids with light-brown centers and pale outer margins. Dorsum has extremely keeled scales. Has loreal pits (between each nostril and eye). Triangular head clearly distinguishable from neck. Robust body. Tail is short and distinct from rest of body. Has rattles at the end of its tail. Terrestrial.

Lachesis melanocephala — Black-headed Bushmaster

RDJ

2.5 m (8 ft)

A large snake. Note dark postocular stripe and black patch on head; also note vertical pupils. Dorsal pattern of dark-brown rhomboids on a light-brown background. Dorsum has extremely keeled, knob-like scales. Has loreal pits (between each nostril and eye). Triangular head clearly distinguishable from neck. Robust body. Tail is short and distinct from rest of body. Terrestrial.

Lachesis stenophrys — Bushmaster

JPM

3 m (10 ft)

Virtually identical to *Lachesis melanocephala* (above), but patch on head is light brown. Terrestrial.

Porthidium nasutum **Rainforest Hognosed Pitviper**

61 cm (2 ft)

Rostral scale elevated, forming the shape of a pointed nose; also note vertical pupils. Pattern on dorsum varies from uniform gray to reddish brown (with small dark squares along body); has 13 to 23 blotches on body. Dorsum has extremely keeled scales. Has loreal pits (between each nostril and eye). Triangular head clearly distinguishable from neck. Robust body. Tail is short and distinct from rest of body. Terrestrial.

JPM

FMC

Porthidium ophryomegas  **Slender Hognosed Pitviper**

80 cm (2.6 ft)

Rostral scale elevated, forming the shape of a pointed nose; note vertical pupils. On light-brown dorsum, also note dark zigzags. Has 21 to 41 blotches on body. Dorsum has extremely keeled scales. Note loreal pits (between each nostril and eye). Triangular head clearly distinguishable from neck. Robust body. Tail is short and distinct from rest of body. Often coiled, with head resting on center top of coil. Terrestrial.

JPM

Porthidium porrasi **White-tailed Hognosed Pitviper**

RDJ

RDJ

60 cm (2 ft)

Rostral scale elevated, forming the shape of a pointed nose; also note vertical pupils. Pattern on dorsum varies from uniform dark brown to pale orangish brown (with small dark squares along body); has 15 to 20 blotches on body. Dorsum has extremely keeled scales. Has loreal pits (between each nostril and eye). Triangular head clearly distinguishable from neck. Robust body. Tail is short and distinct from rest of body. Terrestrial.

Porthidium volcanicum **Río Volcán Hognosed Pitviper**

RDJ (RDJ)

54 cm (21 in)

Rostral scale elevated, forming the shape of a pointed nose; also note vertical pupils. On light-brown dorsum, note dark zigzags. Has 22 to 24 blotches on body. Dorsum has extremely keeled scales. Has loreal pits (between each nostril and eye). Triangular head clearly distinguishable from neck. Robust body. Tail is short and distinct from rest of body. Terrestrial.

Elapidae (6 species)
Coral Snakes & Sea Snakes

Venomous coral snakes are the New World counterpart of the cobras, mambas, and kraits. Their deadly venom (delivered by short, hollow, erect fangs) is neurotoxic, destroying nerve cells, which, in turn, can stop respiration. The distinctive coloration of these snakes warns off potential predators, with apparent great effect: in Costa Rica, 20 nonvenemous colubrid species mimic the coral snakes to obtain similar protective benefits. Venomous coral snakes are secretive and elusive, resulting in a very low rate of human envenomations. They are usually found under leaf litter, both during the day and at night. Another member of this family, the Pelagic Sea Snake (p. 159), lives in marine waters from Mexico to Peru—and in many other bodies of tropical water.

Micrurus alleni **Allen's Coral Snake**

91 cm (3 ft)

Red, black, and yellow rings encircle the body; yellow rings are the narrowest and always fall between red and black rings. Small head; posterior portion of head covered by first yellow ring. Minute black eyes are always located within the region defined by black head cap. Head cap has a dorsal point that intersects first yellow ring. Terrestrial.

RDJ

Micrurus clarki — Clark's Coral Snake

RDJ

71 cm (2.3 ft)

Virtually identical to *Micrurus alleni* (p. 157), but note that the lateral scales of the first yellow ring are bordered with black. Like *M. alleni*, the head cap of *M. clarki* has a dorsal point that intersects first yellow ring, but in this species the head cap doesn't touch the mouth. Terrestrial.

Micrurus mosquitensis — Costa Rican Coral Snake

RDJ

1 m (3.3 ft)

Rings in 3 colors encircle the body: red, black, and yellow. Yellow rings are always the narrowest and always fall between red and black rings. Small head; posterior portion of head covered by first yellow ring. Minute black eyes are always located within the region defined by black head cap. Black head cap does not intersect first yellow ring as it does in *M. alleni* (p. 157) and *M. clarki* (above). Terrestrial.

Micrurus multifasciatus (mipartitus) **Many-banded Coral Snake**

1.1 m (3.5 ft)

RDJ

In spite of its English common name, this snake bears rings not bands. Rings in 2 colors (and of equal width) encircle the body. Black rings occur in all individuals; the second color is orange, red, or white, depending on the individual. Small head; posterior portion of head covered by first orange (or red or white) ring. Minute black eyes are always located within the region defined by black head cap. Terrestrial.

Micrurus nigrocinctus **Central American Coral Snake**

1 m (3.3 ft)

RDJ

Virtually identical to *Micrurus mosquitensis* (p. 158), but it has wider red rings and has white rings in lieu of yellow. Terrestrial.

Pelamis platurus **Pelagic Sea Snake**

91 cm (3 ft)

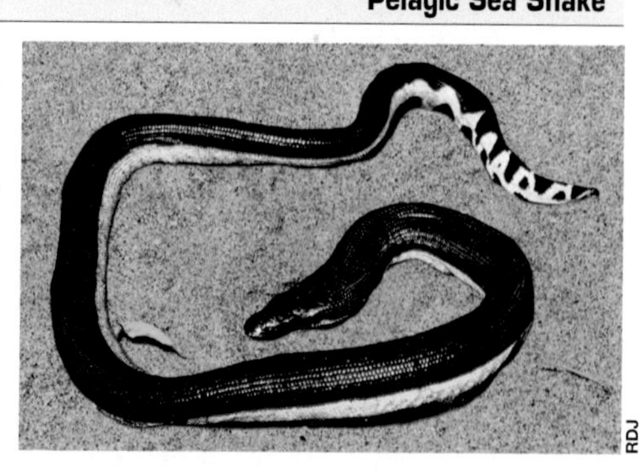

RDJ

Combination of black dorsum and bright yellow venter is distinctive. Tail laterally flattened with a combination of black and yellow blotches. Generally found only in the sea; weak and dying individuals are sometimes washed up onto beaches.

CROCODYLIA (2 species) **Crocodilians**

Alligatoridae (1 species)
Caimans

Caimans belong to the same family as the alligators. On caimans, the longest tooth is the fourth maxillary (fifth on crocodiles). They feed on a wide variety of animals and on carrion. Though mostly aquatic, they are often observed basking on riverbanks. Caimans are generally too small in Costa Rica to pose a threat to humans.

Caiman crocodilus **Spectacled Caiman**

RDJ

2.1 m (7 ft)

Identifying characteristic is an elevated ridge between the eyes. Snout (short and round) is only twice as long as it is wide. Brown to yellowish-green dorsum. Generally found in or near freshwater.

Crocodylidae (1 species)
Crocodiles

On crocodiles, the longest tooth is the fifth maxillary. They often bask on the banks of large rivers. Crocodiles are endangered due to overhunting for their skins. Large crocodiles have been known to attack and kill humans in Costa Rica.

Crocodylus acutus **American Crocodile**

RDJ

4.9 m (16 ft)

Lacks ridge between eyes. Snout (long and pointed) is 3 times longer than it is wide. Dorsum coloration varies from grayish green to olive green; black patches on dorsum change shape toward tail, becoming more band-like. Occurs in both freshwater and brackish water; occasionally seen in sea, near river mouths.

Glossary

A

annuli: Grooved rings that encircle the body of caecilians.

axilla: Synonym for armpit.

B

barbels: Fleshy protuberances on the chin of turtles.

C

carapace: Upper portion of the shell of a turtle.

caudal crest: Elevated crest on some lizards that extends from the top of the neck down the center of the back.

cycloid: Used to describe scales with a semicircular shape.

D

dewlap: Expandable fold of skin on the throat of male anoles that displays one or more color.

digital pads: Expanded pads on the fingers and toes of geckos that enable them to cling to vertical surfaces.

direct development: Development from egg to adult form with no larval stage.

distal: Referring to the end portion of an appendage (as in the *distal* half of legs).

dorsal: Referring to the upper portion of the body.

dorsolateral folds: Folds of skin that extend from the top of body down its sides.

dorsum: Upper portion of the body.

F

fossorial: Living mostly underground or burrowing in leaf litter.

frontal scale: On snakes, the large scale at the center of the top of the head.

H

humeral hook: Fleshy protuberance on the upper arm of the Emerald Glass Frog (*Espadarana prosoblepon*, p. 54).

I

infralabials: Scales on the lower portion of the mouth of lizards and snakes.

internasal scale: On snakes, scale on the top of the head and between the nostrils.

K

keel: Elevated ridge on the upper shell of a turtle.

L

labial pit: On boas, cavity between the lip scales that contains a heat-sensing organ.

labial scales: Scales bordering the mouth of lizards and snakes.

larval stage: In amphibians, stage between egg and adult, i.e., a tadpole.

lateral groove: On salamanders, vertical groove on the side of the body.

lateral tubercles: Tubercles on the side of the body.

loreal pit: On pitvipers, cavity between eye and nostril that contains a heat-sensing organ.

loreal scale: Scale that lies between the postnasal scale (behind nostril) and the preocular scale (in front of the eye).

M

maxillary teeth: Teeth of the upper jaw.

mental scale: Large scale at the tip of the chin.

N

naso-labial groove: On some salamanders, groove that runs from the nostril toward the lip.

nuchal band: A band across the neck or nape.

O

oviparous: Egg laying.

P

parotid gland: On most toads, gland behind the eye that contains toxins.

parthenogenetic: Asexual reproduction that produces offspring that are all female clones of the mother.

pericardium: Tissue surrounding the heart.

phylogenetics: The study of evolutionary relatedness through DNA analysis.

plastron: That section of a turtle's shell that protects its underside.

postocular: Referring to a region behind the eye.

prefrontal scale: Scale that lies just in front of the frontal scale.

premontane rainforest: Middle-elevation rainforest, typically from 500 to 1,500 meters (1,640 to 4,920 ft).

prenasal scale: Scale located just before the nostril.

preocular scale: Scale located in front of the eye.

primary annuli: In caecilians, grooves that encircle the body.

R

rostral scale: Large scale at the tip of the snout.

rugose: Refers to an uneven or rough appearance, often used to describe skin with bumps or warts.

S

secondary annuli: In caecilians, grooves that lie between the primary annuli; secondary annuli do not encircle the entire body (unlike primary annuli).

shields: Hardened epidermal plates that make up the shells of turtles. Also found on the heads, necks, and flippers of hard shelled sea turtles.

subarticular tubercle: Tubercle located below the digit joint.

submental scales: Large scales that lie posterior to the mental scale.

supernumerary tubercle: Tubercle located below the digit.

supralabial scale: Scale on the upper portion of the mouth of lizards and snakes.

systematics: Study of the classification of life forms.

T

tarsal: Pertaining to the ankle.

tubercle: Any raised area such as a wart.

tympanic membrane: Membrane that covers the ear opening.

tympanum: The ear, covered by a membrane.

V

vent: Anal opening.

venter: The lower surface of an animal.

vertebral stripe: A stripe that runs lengthwise, down the center of the back.

viviparous: Giving birth to fully formed juveniles.

Acknowledgments

The authors thank a number of people whose contributions helped make this publication possible: William Becker (Central States Serpentarium); Federico Bolaños (Universidad de Costa Rica); Lance Brauchman; Minor Camacho (Parque Viborana); Vilma E. Castillo, Esteban Rojas, and Rosario Gómez (Tortufauna); Alan Day; Quetzal Dwyer (Reptilandia); Micky Finke; Marcel Goldmann and Robert Meindinger (World of Snakes); Jeff Hardwick; Weindert Hensen; Victor Hugo Quesada (Serpentario El Castillo); William W. Lamar (GreenTracks); Joseph P. Marek Jr.; George Neth; John Nichols; Marcela Rojas Vega; Mahmood Sasa (Instituto Clodomiro Picado); Alejandro Solórzano (Serpentario Nacional); Rodolfo Vargas Leitón and Ricardo Sanabria Mora (Refugio Herpetológico de Costa Rica); and Andrés Vega. The authors also extend their gratitude to several organizations—and their staff—that offered help along the way: Frog Pond of Monteverde, Serpentario Monteverde, and La Paz Waterfall Gardens.

Finally, this book would not have been possible without the photographs and illustrations generously provided by the following people:

RDJ	Richard Dennis Johnston
FMC	Federico Muñoz Chacón
AV	Andrés Vega
TL	Twan Leenders
JPM	Joseph P. Marek Jr.
JCN	Jonathan Chacón Navarro
DH	David Hulmes
RVL	Rodolfo Vargas Leitón
WB	William Becker
RP	Robert Puschendorf
LB	Lance Brauckman
GF	Giovanni Fagioli
MRV	María Marcela Rojas Vega
GS	Gerald Sylvester
JES	John E. Simmons
JCS	Javier Carazo Salazar
JF	Jerry Fife
RG	Russ Gurley
WH	Wiendert Hensen
GSM	Gustavo Serrano Mora
SS	Scott Solar
JVH	Jeff V. Hardwick
SY	Susan Yow

The photograph of *Dendropsophus ebraccatus* on page 1 was provided by Richard Dennis Johnston. The photograph of *Anolis biporcatus* on page 66 was provided by Greg Basco.

Systematic Index

Agalychnis annae 36
Agalychnis calcarifer, see *Cruziohyla* 39
Agalychnis callidryas 37
Agalychnis lemur 37
Agalychnis saltator 38
Agalychnis spurrelli 38
Agkistrodon bilineatus 148
Alligatoridae 160
Allobates talamancae 61
Amastridium veliferum 112
Ameiva ameiva 98
Ameiva festiva 99
Ameiva leptophrys 99
Ameiva quadrilineata 99
Ameiva undulata 100
Amphibia 1
Anadia ocellata 101
Anguidae 103
Anolis altae 82
Anolis aquaticus 83
Anolis biporcatus 83
Anolis capito 83
Anolis carpenteri 84
Anolis cristatellus 84
Anolis cupreus 84
Anolis frenatus 85
Anolis fungosus 85
Anolis humilis 85
Anolis ibanezi 85
Anolis insignis 86
Anolis intermedius 86
Anolis kemptoni 87
Anolis lemurinus 87
Anolis limifrons 87
Anolis lionotus 88
Anolis microtus 88
Anolis pachypus 88
Anolis pentaprion 88
Anolis polylepis 89
Anolis townsendi 89
Anolis tropidolepis 89
Anolis unilobatus 90
Anolis vociferans 90
Anolis woodi 90
Anomalepididae 106
Anomalepis mexicanus 106
Anotheca spinosa 38
Anura 14
Aromobatidae 61
Aspidoscelis, see *Cnemidophorus* 100
Atelopus chiriquiensis 14
Atelopus chirripoensis 15
Atelopus senex 15
Atelopus varius 15
Atropoides mexicanus 149
Atropoides picadoi 149
Bachia blairi 101
Basiliscus basiliscus 76
Basiliscus plumifrons 77
Basiliscus vittatus 77
Bataguridae, see Geoemydidae 74
Boa constrictor 109
Boidae 109
Bolitoglossa alvaradoi 4
Bolitoglossa bramei 4
Bolitoglossa cerroensis 5
Bolitoglossa colonnea 5
Bolitoglossa compacta 5
Bolitoglossa diminuta 5
Bolitoglossa epimela 6
Bolitoglossa gomezi 6
Bolitoglossa gracilis 6
Bolitoglossa lignicolor 6
Bolitoglossa marmorea 6
Bolitoglossa minutula 7
Bolitoglossa nigrescens 7
Bolitoglossa obscura 7
Bolitoglossa pesrubra 7
Bolitoglossa robusta 8
Bolitoglossa schizodactyla 8
Bolitoglossa sombra 8
Bolitoglossa sooyorum 8
Bolitoglossa striatula 9
Bolitoglossa subpalmata 9
Bolitoglossa tica 9
Bothriechis lateralis 150

Bothriechis nigroviridis 150
Bothriechis schlegelii 151
Bothriechis supraciliaris 151
Bothrops asper 152
Bufo, see *Incilius* 16
Bufo haematiticus, see *Rhaebo* 20
Bufo marinus, see *Rhinella* 20
Bufonidae 14
Caeciliiadae 2
Caiman crocodilus 160
Caretta caretta 67
Caudata 4
Celestus cyanochlorus 103
Celestus hylaius 104
Celestus orobius 104
Centrolene prosoblepon,
see *Espadarana* 54
Centrolene ilex, see *Sachatamia* 56
Centrolenidae 53
Cerrophidion godmani 153
Chelonia mydas 68
Cheloniidae 67
Chelydra acutirostris 70
Chelydridae 70
Chironius carinatus 112
Chironius exoletus 113
Chironius grandisquamis 113
Clelia clelia 114
Clelia scytalina 114
Cnemidophorus deppii 100
Cochranella albomaculata,
see *Sachatamia* 56
Cochranella euknemos 53
Cochranella granulosa 53
Cochranella pulverata, see *Teratohyla* 57
Cochranella spinosa, see *Teratohyla* 57
Colostethus, see *Silverstoneia* 60
Colostethus talamancae,
see *Allobates* 61
Coleonyx mitratus 92
Coloptychon rhombifer 104
Coluber, see *Masticophis* 128
Colubridae 112
Coniophanes bipunctatus 115
Coniophanes fissidens 115
Coniophanes piceivittis 115
Conophis lineatus 116
Corallus annulatus 109
Corallus ruschenbergerii 110
Corytophanes cristatus 78
Corytophanidae 76
Craugastor andi 23
Craugastor angelicus 23
Craugastor bransfordii 23
Craugastor catalinae 23
Craugastor crassidigitus 24
Craugastor cuaquero 24
Craugastor escoces 24
Craugastor fitzingeri 24
Craugastor fleischmanni 25
Craugastor gollmeri 25
Craugastor gulosus 25
Craugastor megacephalus 25
Craugastor melanostictus 26
Craugastor mimus 26
Craugastor noblei 26
Craugastor obesus 27
Craugastor persimilis 27
Craugastor phasma 27
Craugastor podiciferus 27
Craugastor polyptychus 28
Craugastor ranoides 28
Craugastor rayo 28
Craugastor rhyacobatrachus 28
Craugastor rugosus 29
Craugastor stejnegerianus 29
Craugastor talamancae 29
Craugastor taurus 30
Craugastor underwoodi 30
Craugastoridae 23
Crepidophryne chompipe 15
Crepidophryne epiotica 16
Crepidophryne guanacaste 16
Cristantophis nevermanni 116
Crocodylia 160
Crocodylidae 160
Crocodylus acutus 160
Crotalus simus 153
Cruziohyla calcarifer 39
Ctenonotus, see *Anolis* 84
Ctenosaura quinquecarinata 79
Ctenosaura similis 79
Dactyloa, see *Anolis* 85, 86, 88
Dactyloidae 82
Dendrobates auratus 58
Dendrobates granulifera,
see *Oophaga* 58
Dendrobates pumilio, see *Oophaga* 59

Dendrobatidae 58
Dendrophidion nuchale 116
Dendrophidion paucicarinatum 117
Dendrophidion percarinatum 117
Dendrophidion vinitor 117
Dendropsophus ebraccatus 39
Dendropsophus microcephalus 39
Dendropsophus phlebodes 40
Dermochelyidae 69
Dermochelys coriacea 69
Dermophis costaricense 2
Dermophis glandulosus 2
Dermophis gracilior 2
Dermophis occidentalis 3
Dermophis parviceps 3
Diasporus diastema 21
Diasporus hylaeformis 21
Diasporus tigrillo 21
Diasporus ventrimaculatus 22
Diasporus vocator 22
Diploglossus bilobatus 105
Diploglossus monotropis 105
Dipsas articulata 118
Dipsas bicolor 118
Dipsas tenuissima 118
Dryadophis, see *Mastigodryas* 129
Drymarchon melanurus 119
Drymobius margaritiferus 119
Drymobius melanotropis 119
Drymobius rhombifer 120
Duellmanohyla lythrodes 40
Duellmanohyla rufioculis 40
Duellmanohyla uranochroa 41
Ecnomiohyla fimbrimembra 41
Ecnomiohyla miliaria 41
Ecnomiohyla sukia 41
Elapidae 157
Eleutherodactylidae 21
Eleutherodactylus altae, see *Pristimantis* 30
Eleutherodactylus andi, see *Craugastor* 23
Eleutherodactylus angelicus, see *Craugastor* 23
Eleutherodactylus bransfordii, see *Craugastor* 23
Eleutherodactylus bufoniformis, see *Strabomantis* 32
Eleutherodactylus caryophyllaceus, see *Pristimantis* 31
Eleutherodactylus catalinae, see *Craugastor* 23
Eleutherodactylus cerasinus, see *Pristimantis* 31
Eleutherodactylus coqui 22
Eleutherodactylus crassidigitus, see *Craugastor* 24
Eleutherodactylus cruentus, see *Pristimantis* 31
Eleutherodactylus cuaquero, see *Craugastor* 24
Eleutherodactylus diastema, see *Diasporus* 21
Eleutherodactylus escoces, see *Craugastor* 24
Eleutherodactylus fitzingeri, see *Craugastor* 24
Eleutherodactylus fleischmanni, see *Craugastor* 25
Eleutherodactylus gaigei, see *Pristimantis* 32
Eleutherodactylus gollmeri, see *Craugastor* 25
Eleutherodactylus gulosus, see *Craugastor* 25
Eleutherodactylus hylaeformis, see *Diasporus* 21
Eleutherodactylus megacephalus, see *Craugastor* 25
Eleutherodactylus melanostictus, see *Craugastor* 26
Eleutherodactylus mimus, see *Craugastor* 26
Eleutherodactylus moro, see *Pristimantis* 32
Eleutherodactylus noblei, see *Craugastor* 26
Eleutherodactylus obesus, see *Craugastor* 27
Eleutherodactylus pardalis, see *Pristimantis* 32
Eleutherodactylus persimilis, see *Craugastor* 27
Eleutherodactylus phasma, see *Craugastor* 27
Eleutherodactylus podiciferus, see *Craugastor* 27
Eleutherodactylus polyptychus, see *Craugastor* 28

Eleutherodactylus ranoides, see *Craugastor* 28
Eleutherodactylus rayo, see *Craugastor* 28
Eleutherodactylus rhyacobatrachus, see *Craugastor* 28
Eleutherodactylus ridens, see *Pristimantis* 32
Eleutherodactylus rugosus, see *Craugastor* 29
Eleutherodactylus stejnegerianus, see *Craugastor* 29
Eleutherodactylus talamancae, see *Craugastor* 29
Eleutherodactylus taurus, see *Craugastor* 30
Eleutherodactylus tigrillo, see *Diasporus* 21
Eleutherodactylus underwoodi, see *Craugastor* 30
Eleutherodactylus vocator, see *Diasporus* 22
Emydidae 73
Engystomops pustulosus 35
Enuliophis sclateri 120
Enulius flavitorques 120
Epictia goudotii 107
Epricrates maurus 110
Eretmochelys imbricata 68
Erythrolamprus bizonus 121
Erythrolamprus mimus 121
Espadarana prosoblepon 54
Eumeces, see *Mesoscincus* 97
Gastrophryne pictiventris 62
Gastrotheca cornuta 36
Geckkonidae 92
Geoemydidae 74
Geophis brachycephalus 121
Geophis downsi 122
Geophis godmani 122
Geophis hoffmanni 122
Geophis ruthveni 122
Geophis talamancae 122
Geophis zeledoni 123
Gonatodes albogularis 92
Gymnophiona 2
Gymnophthalmus speciosus 102
Gymnophthalmidae 101
Gymnopis multiplicata 3
Helminthophis frontalis 106
Hemidactylus frenatus 93
Hemidactylus garnoti 93
Hemiphractidae 36
Hyalinobatrachium chirripoi 54
Hyalinobatrachium colymbiphyllum 54
Hyalinobatrachium fleischmanni 55
Hyalinobatrachium talamancae 55
Hyalinobatrachium valerioi 55
Hyalinobatrachium vireovittatum 56
Hydromorphus concolor 123
Hyla angustilineata, see *Isthmohyla* 43
Hyla calypsa, see *Isthmohyla* 43
Hyla colymba, see *Hyloscirtus* 42
Hyla debilis, see *Isthmohyla* 44
Hyla ebraccata, see *Dendropsophus* 39
Hyla fimbrimembra, see *Ecnomiohyla* 41
Hyla lancasteri, see *Isthmohyla* 44
Hyla legleri, see *Ptychohyla* 47
Hyla loquax, see *Tlalocohyla* 52
Hyla microcephala, see *Dendropsophus* 39
Hyla miliaria, see *Ecnomiohyla* 41
Hyla palmeri, see *Hyloscirtus* 42
Hyla phlebodes, see *Dendropsophus* 40
Hyla picadoi, see *Isthmohyla* 44
Hyla pictipes, see *Isthmohyla* 44
Hyla pseudopuma, see *Isthmohyla* 45
Hyla rivularis, see *Isthmohyla* 45
Hyla rosenbergi, see *Hypsiboas* 42
Hyla rufitela, see *Hypsiboas* 43
Hyla sukia, see Ecnomiohyla 41
Hyla tica, see *Isthmohyla* 46
Hyla xanthosticta, see *Isthmohyla* 46
Hyla zeteki, see *Isthmohyla* 46
Hylidae 36
Hylomantis lemur, see *Agalychnis* 37
Hyloscirtus colymba 42
Hyloscirtus palmeri 42
Hypopachus variolosus 62
Hypsiboas rosenbergi 42
Hypsiboas rufitelus 43
Iguana iguana 80
Iguanidae 79
Imantodes cenchoa 123
Imantodes gemnistratus 124
Imantodes inornatus 124
Incilius aucoinae 16
Incilius coccifer 17
Incilius coniferus 17
Incilius fastidosus 17

Incilius holdridgei 18
Incilius luetkenii 18
Incilius melanochlorus 19
Incilius periglenes 19
Incilius valliceps 20
Isthmohyla angustilineata 43
Isthmohyla calypsa 43
Isthmohyla debilis 44
Isthmohyla lancasteri 44
Isthmohyla picadoi 44
Isthmohyla pictipes 44
Isthmohyla pseudopuma 45
Isthmohyla rivularis 45
Isthmohyla tica 46
Isthmohyla xanthosticta 46
Isthmohyla zeteki 46
Kinosternidae 71
Kinosternon angustipons 71
Kinosternon leucostomum 71
Kinosternon scorpioides 72
Lachesis melanocephala 154
Lachesis stenophrys 154
Lampropeltis triangulum 124
Leiuperidae 35
Lepidoblepharis xanthostigma 93
Lepidochelys olivacae 69
Lepidodactylus lugubris 94
Lepidophyma flavimaculatum 96
Lepidophyma reticulatum 97
Leposoma southi 102
Leptodactylidae 33
Leptodactylus bolivianus 33
Leptodactylus fragilis 33
Leptodactylus melanonotus 34
Leptodactylus poecilochilus 34
Leptodactylus savageii 35
Leptodeira annulata 125
Leptodeira nigrofasiata 125
Leptodeira rubricata 125
Leptodeira septentrionalis 126
Leptodrymus pulcherrimus 126
Leptophis ahaetulla 126
Leptophis depressirostris 127
Leptophis mexicanus 127
Leptophis nebulosus 127
Leptophis riveti 128
Leptotyphlopidae 107
Leptotyphlops, see *Epictia* 107
Liophis epinephalus 128
Liotyphlops albirostris 106
Lithobates forreri 63
Lithobates taylori 64
Lithobates vaillanti 64
Lithobates vibicarius 64
Lithobates warszewitschii 65
Loxocemidae 108
Loxocemus bicolor 108
Mabuya unimarginata 97
Masticophis mentovarius 128
Mastigodryas melanolomus 129
Mesaspis monticola 105
Mesoscincus managuae 97
Microhylidae 62
Micrurus alleni 157
Micrurus clarki 158
Micrurus mosquitensis 158
Micrurus multifasciatus 159
Micrurus nigrocinctus 159
Nelsonophryne aterrima 63
Neusticurus, see *Potamites* 102
Ninia celata 129
Ninia maculata 129
Ninia psephota 130
Ninia sebae 130
Norops, see *Anolis* 82
Nothopsis rugosus 130
Nototriton abscondens 9
Nototriton gamezi 10
Nototriton guanacaste 10
Nototriton major 10
Nototriton picadoi 10
Nototriton richardi 10
Nototriton tapanti 11
Oedipina alfaroi 11
Oedipina alleni 11
Oedipina altura 11
Oedipina carablanca 11
Oedipina collaris 11
Oedipina cyclocauda 12
Oedipina gracilis 12
Oedipina grandis 12
Oedipina pacificensis 12
Oedipina paucidentata 12
Oedipina poelzi 13
Oedipina pseudouniformis 13
Oedipina savagei 13
Oedipina uniformis 13
Oophaga granulifera 58
Oophaga pumilio 59
Oscaecilia osae 3

Osteopilus septentrionalis 47
Oxybelis aeneus 131
Oxybelis brevirostris 131
Oxybelis fulgidus 131
Oxyrhopus petola 132
Pelamis platurus 159
Plethodontidae 4
Phrynohyas venulosa,
see *Trachycephalus* 52
Phrynosomatidae 81
Phyllobates lugubris 59
Phyllobates vittatus 60
Phyllodactylus tuberculosus 94
Phyllomedusa lemur, see *Agalychnis* 37
Physalaemus pustulosus,
see *Engystomops* 35
Pliocercus euryzonus 132
Polychrotidae 91
Polychrus gutturosus 91
Porthidium nasutum 155
Porthidium ophryomegas 155
Porthidium porrasi 156
Porthidium volcanicum 156
Potamites apodemus 102
Pristimantis altae 30
Pristimantis caryophyllaceus 31
Pristimantis cerasinus 31
Pristimantis cruentus 31
Pristimantis gaigei 32
Pristimantis moro 32
Pristimantis pardalis 32
Pristimantis ridens 32
Pseustes poecilonotus 132
Ptychoglossus plicatus 103
Ptychohyla legleri 47
Rana, see *Lithobates* 63
Ranidae 63
Reptilia 66
Rhadinaea calligaster 133
Rhadinaea decorata 133
Rhadinaea pulveriventris 133
Rhadinella godmani 134
Rhadinella serperaster 134
Rhaebo haematiticus 20
Rhinella marina 20
Rhinobothryum bovallii 134
Rhinoclemmys annulata 74
Rhinoclemmys funerea 75
Rhinoclemmys pulcherrima 75
Rhinophrynidae 14
Rhinophrynus dorsalis 14
Sachatamia albomaculata 56
Sachatamia ilex 56
Sauria 76
Scaphiodontophis annulatus 135
Sceloporus malachiticus 81
Sceloporus squamosus 81
Sceloporus variabilis 82
Scinax boulengeri 48
Scinax elaeochrous 48
Scinax staufferi 49
Scincidae 97
Scolecophis atrocinctus 135
Senticolis triaspis 135
Serpentes 106
Sibon annulatus 136
Sibon anthracops 136
Sibon argus 136
Sibon dimidiatus 137
Sibon lamari 137
Sibon longifrenis 138
Sibon nebulatus 138
Silverstoneia flotator 60
Silverstoneia nubicola 61
Siphlophis compressus 138
Smilisca baudinii 49
Smilisca phaeota 50
Smilisca puma 50
Smilisca sila 51
Smilisca sordida 51
Sphaerodactylus graptolaemus 94
Sphaerodactylus homolepis 95
Sphaerodactylus millepunctatus 95
Sphaerodactylus pacificus 95
Sphenomorphus cherriei 98
Spilotes pullatus 139
Squamata 76, 106
Stenorrhina degenhardtii 139
Stenorrhina freminvillii 140
Strabomantidae 30
Strabomantis bufoniformis 32
Tantilla alticola 140
Tantilla armillata 141
Tantilla reticulata 141
Tantilla ruficeps 141
Tantilla schistosa 142
Tantilla supracincta 142
Tantilla vermiformis 142

Teiidae 98
Teratohyla pulverata 57
Teratohyla spinosa 57
Testudines 67
Thamnophis marcianus 143
Thamnophis proximus 143
Thecadactylus rapicauda 96
Tlalocohyla loquax 52
Trachemys emolli 73
Trachemys venusta 73
Trachycephalus venulosus 52
Tretanorhinus nigroluteus 143
Trimetopon gracile 144
Trimetopon pliolepis 144
Trimetopon simile 144
Trimetopon slevini 144
Trimetopon viquezi 144
Trimorphodon quadruplex 145
Tripanurgos, see *Siphlophis* 138
Tropidodipsas sartorii 145
Tropidophiidae, see Ungaliophiidae 111
Typhlopidae 107
Typhlops costaricensis 107
Ungaliophiidae 111
Ungaliophis panamensis 111
Urotheca decipiens 145
Urotheca euryzona, see *Pliocercus* 132
Urotheca fulviceps 146
Urotheca guentheri 146
Urotheca myersi 146
Urotheca pachyura 146
Viperidae 148
Xantusiidae 96
Xenodon rabdocephalus 147

About the Authors

Federico Muñoz Chacón is a Costa Rican research biologist and science educator. Inspired by professor Douglas C. Robinson, he has dedicated his life to studying the reptiles and amphibians of Costa Rica and to promoting ecological awareness. Federico also leads expeditions to rainforests in a number of Latin American countries. He lives in Monteverde, Costa Rica, where he owns and operates the biological reserve Terra Viva.

Native Californian **Richard Dennis Johnston** has been studying and photographing reptiles and amphibians for more than 45 years. Since 1997 he has focused his research and photography work on Costa Rican herpetofauna. When not in Costa Rica, he works as a biochemist in Newark, Delaware, and serves as a member of the advisory board of the International Herpetological Symposium.